The Memory System

You Can't Avoid It,
You Can't Ignore It,
You Can't Fake It

Synthesis Lectures on Computer Architecture

Editor
Mark D. Hill, *University of Wisconsin, Madison*

Synthesis Lectures on Computer Architecture publishes 50 to 150 page publications on topics pertaining to the science and art of designing, analyzing, selecting and interconnecting hardware components to create computers that meet functional, performance and cost goals.

The Memory System: You Can't Avoid It, You Can't Ignore It, You Can't Fake It
Bruce Jacob
2009

Fault Tolerant Computer Architecture
Daniel J. Sorin
2009

The Datacenter as a Computer: An Introduction to the Design of Warehouse-Scale Machines
Luiz André Barroso, Urs Hölzle
2009

Computer Architecture Techniques for Power-Efficiency
Stefanos Kaxiras, Margaret Martonosi
2008

Chip Multiprocessor Architecture: Techniques to Improve Throughput and Latency
Kunle Olukotun, Lance Hammond, James Laudon
2007

Transactional Memory
James R. Larus, Ravi Rajwar
2006

Quantum Computing for Computer Architects
Tzvetan S. Metodi, Frederic T. Chong
2006

The Memory System: You Can't Avoid It, You Can't Ignore It, You Can't Fake It

Bruce Jacob

ISBN: 978-3-031-00596-1   paperback
ISBN: 978-3-031-01724-7   ebook

DOI 10.1007/978-3-031-01724-7

A Publication in the Springer series
SYNTHESIS LECTURES ON COMPUTER ARCHITECTURE

Lecture #7
Series Editor: Mark D. Hill, *University of Wisconsin, Madison*

Series ISSN
Synthesis Lectures on Computer Architecture
Print 1935-3235   Electronic 1935-3243

The Memory System

You Can't Avoid It,
You Can't Ignore It,
You Can't Fake It

Bruce Jacob
University of Maryland

with contributions by

Sadagopan Srinivasan
Intel

David T. Wang
MetaRAM

ABSTRACT

Today, computer-system optimization, at both the hardware and software levels, must consider the details of the memory system in its analysis; failing to do so yields systems that are increasingly inefficient as those systems become more complex. This lecture seeks to introduce the reader to the most important details of the memory system; it targets both computer scientists and computer engineers in industry and in academia. Roughly speaking, computer scientists are the users of the memory system, and computer engineers are the designers of the memory system. Both can benefit tremendously from a basic understanding of how the memory system really works: the computer scientist will be better equipped to create algorithms that perform well, and the computer engineer will be better equipped to design systems that approach the optimal, given the resource limitations. Currently, there is consensus among architecture researchers that the memory system is "the bottleneck," and this consensus has held for over a decade. Somewhat inexplicably, most of the research in the field is still directed toward improving the CPU to better tolerate a slow memory system, as opposed to addressing the weaknesses of the memory system directly. This lecture should get the bulk of the computer science and computer engineering population up the steep part of the learning curve. Not every CS/CE researcher/developer needs to do work in the memory system, but, just as a carpenter can do his job more efficiently if he knows a little of architecture, and an architect can do his job more efficiently if he knows a little of carpentry, giving the CS/CE worlds better intuition about the memory system should help them build better systems, both software and hardware.

KEYWORDS

memory systems, storage systems, memory scheduling, system simulation, memory-system design, prefetching, hash-associative cache, virtual memory, superpages, memory power, storage power, translation lookaside buffers, cache design, DRAM systems, memory bandwidth, memory latency, memory trends

Contents

Prelude: Why Should I Care About the Memory System?

(you cannot possibly get good performance by ignoring it)

Whether you like to admit it or not, your code does not execute in the Æther: to run, programs must use physical resources, for both computation and storage. While it is certainly possible to write code with absolutely no knowledge of what those physical resources are, nor how they work, doing so is unlikely to yield behavior that represents either good performance or minimal energy consumption. This lesson was learned long ago in the context of computational resources; as a result, today any good analysis for hardware or software design considers the specifics of the processor/s (and the multiprocessor interconnect, if applicable) in great detail.

We have reached a point at which the lesson must also be learned in the context of the memory system. Today, computer-system optimization, at both the hardware and software levels, must consider the details of the memory system in its analysis; failing to do so yields systems that are increasingly inefficient as those systems become more complex. Why this should be true is intuitively clear: as systems become more complex, hitherto unimportant interactions between components begin to appear and even dominate. Models that neglect these interactions become increasingly inaccurate as the systems they purport to represent become more complex. One consequence is that researchers and developers who rely on those models run increasing risks of building systems that fail to deliver promised performance or exceed their expected power and cooling requirements. Previous systems were relatively simple, and so simple models sufficed; today, systems are no longer at all simple, either on the computation side or on the storage side. The simple conclusion is that simple models no longer suffice.

This lecture seeks to introduce the reader to the most important details of the memory system; it targets both software and hardware designers, and therefore the range of topics discussed is wide. The obvious question is *what do I need to know about the memory system?*

- You need to know the basics. This includes the way your code interacts with physical storage, fundamentals (not technical details) of DRAM-system operation, and some insights into present-day technologies and trends.

- You need to know how badly you can be misled by simplistic models. Nearly every simulator available today uses a simplistic model of the memory system (including many that purport to use a "cycle-accurate model of the memory system"). Nearly every study uses these models—including studies in both academia and industry. An increasing trend is that software and hardware mechanisms described in the literature make it into production systems but are disabled afterward. These issues are all related.

- You need to know what the near future looks like, as the details are likely to change. A handful of important problems face the community, and the problems are not going away. Some are show-stoppers—if not solved soon, performance will flatline (many would argue this has already happened). The interesting thing is that many of the problems have common solutions, which suggests possible future directions. Whether any of this comes true is less important than being aware of the problems, considering the possible solutions, and being prepared for things to change, whichever direction the community takes.

As mentioned, the lecture targets both computer scientists and computer engineers, both in industry and in academia. Roughly speaking, computer scientists are the users of the memory system, and computer engineers are the designers of the memory system. Both can benefit tremendously from a basic understanding of how the memory system really works: the computer scientist will be better equipped to create algorithms that perform well (as we all know, big-O notation is only a first-order understanding, since no algorithm works in a vacuum), and the computer engineer will be better equipped to design systems that approach the optimal, given the resource limitations.

Currently, there is consensus among architecture researchers that the memory system is "the bottleneck," and this consensus has held for over a decade. Somewhat inexplicably, most of the research in the field is still directed toward improving the CPU to better tolerate a slow memory system, as opposed to addressing the weaknesses of the memory system directly. This lecture should get the bulk of the computer science and computer engineering population up the steep part of the learning curve. Not every CS/CE researcher/developer needs to do work in the memory system, but, just as a carpenter can do his job more efficiently if he knows a little of architecture, and an architect can do his job more efficiently if he knows a little of carpentry, giving the CS/CE worlds better intuition about the memory system should help them build better systems, both software and hardware.

CHAPTER 1

Primers

(you can't fake it; do it right or don't bother doing it at all)

We begin with a few overviews of the memory system. The first is a look at how data is stored in memory: knowledge of data layout is critical in all compiler work, and many programming languages indirectly expose the programmer to data layout as well (consider, for instance, the difference between `char *str;` and `char str[];` in C). The second overview presents a brief tutorial on how modern DRAM systems work. The last overview puts recent developments in memory systems in perspective, to give context for the points made in the following chapters.

1.1 YOUR CODE DOES NOT RUN IN A VACUUM

This primer describes how your code and data are represented in the physical memory system, using C as an example. This section can be skipped by anyone who already knows this stuff, as the rest of the lecture is relatively independent. However, faculty who already know this stuff may wish nonetheless to read through it and make sure that you teach (emphasize) the material in your programming classes. I have seen numerous self-proclaimed coding wizards (both students and industry practitioners alike) who had no idea what the memory system was nor how it worked, and, for instance, they would be dumbfounded when their code started dripping errors when ported to a new compiler that, counter to the modern "kinder, gentler" movement in idiot-proof compiler design, failed to automatically set all uninitialized pointers to NULL on behalf of the programmer.

1.1.1 DATA AND ITS REPRESENTATION

First and foremost, you need to understand that there is this thing called *memory* that holds all of your variables. C allows you to declare data types, such as the following:

```
int a, b, c;
int *ip;
char str[] = "strlen 18, size 19";
char *cp;
int A[2] = { 5551212, 5551313 };
double f;
```

These data items are not simply floating around somewhere; they are stored in the machine's memory. You refer to them with nice names such as `str` and `cp`, but the machine does not know these names; they are constructions of the compiler and are translated into numeric addresses for the sake of the machine.

For instance, if we are talking about a 32-bit machine and assume that the data are all placed in the same area of memory (sometimes strings go somewhere other than integer and floating-point data), the memory area would look a little like the following (the string `str` and the array `A` are the only data items initialized, so they are the only items containing any information):

starting addr		< —— 32-bit quantity —— >		
52				
48	f:			
44		5551313		
40	A:	5551212		
36	cp:	18		
32		1	9	\0	?
28		i	z	e	
24		8	,		s
20		e	n		1
16	str:	s	t	r	1
12	ip:				
8	c:				
4	b:				
0	a:				

This example assumes that we will give locations in memory the following addresses:

< —— 32-bit quantity —— >			
etc	etc	...	...
8	9	10	11
4	5	6	7
0	1	2	3

So, the first three data items in memory (starting with item a) are four-byte quantities interpreted as integers. The compiler makes sure that the machine interprets the data correctly. Therefore, when we say:

```
a = b;
```

this translates to:

```
copy four bytes, starting at memory location b,
to memory location a
```

When we say:

```
str[0] = str[7];
```

this translates to:

```
copy one byte, starting at memory location str+7,
to memory location str+0
```

And when we say:

```
f = 3.14159;
```

this translates to:

```
place into memory location f an 8-byte quantity
having the value 3.14159
```

Note that the **machine** does no such interpreting; if you told it to jump to address 0 (or whatever address at which we have loaded item a), it would interpret the data as an instruction. If you told it to load the data there and print it out as a string, it would treat the data as a bunch of characters. If you told it to perform floating-point operations on it, the machine would treat that data as a floating-point number. Nonetheless, our C compiler will (attempt to) keep all usage straight and will make sure that we only use the 4-byte data item at location a as an integer.

At memory location ip we have something slightly different. This piece of data is declared as a pointer to an integer. This means that the data item will be used to reference integers. It will not hold integer values itself; it will merely point to storage locations that hold integers.

The distinction is critical; we will get to it in detail in a moment. Let us suppose that both a and ip contain the value 8. Then the following lines of code do several different things:

```
a = 8;
a = 4;
ip = 8;
*ip = 4;
```

The first line places the value 8 into the location a, overwriting whatever was previously there. The second line places the value 4 into the location a, overwriting the value 8. The third line places the value 8 into the location ip, overwriting whatever was previously there. The fourth line de-references the pointer ip and says to the computer: place the value 4 into the integer referenced by ip. Since ip currently has the value 8, the computer places the value 4 at the location 8, which happens to be item c.

The next item in the list is the character array str, which has been initialized to the value "strlen 18, size 19". The string-length of the character array is 18 bytes, but the amount of space that it occupies is 19 bytes; strings are null-terminated, so for every string there is an extra zero-byte at the end. The amount of space the compiler allocates to hold the string is actually 20 bytes—a multiple of 4 (the word size in our example). We will get to this in more detail in a moment.

Note that the first character in a string starts at the lowest memory address, and the string works its way upward in memory, not downward. We could use the character pointer cp to point at the data elements in this character array. Say we do the following:

```
cp = str;
cp++;
cp = cp + 1;
*cp = '\0';
```

The first line assigns location of str to the variable cp, therefore the value 16 gets put into the location marked cp. The next line increments cp, so the pointer gets the new value 17, and so points at the second element in str (the 't' character). The third line does the same, and cp points to the 'r' character. The fourth line does not change the value of cp; it changes the value of the thing that cp points to: the 'r' character. At this point, we have changed the data in the string. We have placed a NULL character where the 'r' used to be. Memory now looks like the following:

starting addr		< −− 32-bit quantity −− >			
52					
48	f:				
44			5551313		
40	A:		5551212		
36	cp:		18		
32		1	9	\0	?
28		i	z	e	
24		8	,		s
20		e	n		1
16	str:	s	t	r	1
12	ip:				
8	c:				
4	b:				
0	a:				

If we were to do a strlen(str), we would get the value 2, because there are now only two bytes in the string before we hit a NULL terminator. However, the rest of the string did not go away; it is still there, occupying space.

Speaking of the rest of the data sitting there, occupying space, what is the '?' character at memory location 35? It is an unknown. The compiler typically allocates space in units of 4 bytes (the fundamental unit in the 32-bit machine: 32 bits). Since our string only required 19 bytes, we had a little space left over, and the compiler did *not* want to start the next item (the character pointer) at an odd address, so the compiler wastes the space in between the end of the string and the beginning of the character pointer, hoping that you will never go looking for it, or accidentally use it. It might contain the value '\0' but it could just as well contain the value 'Z' or (more likely) something outside the ASCII range, like 129 or 231 or 17. You cannot count on this value, but you can generally count on the space being there.

After the character pointer comes an integer array A. It has two elements, 5551212 and 5551313. Let us look at a variant of earlier code:

```
ip = A;
ip++;
ip = ip + 1;
*ip = 0;
```

The first line points ip to the array A; at this point, the variable ip contains the value 40. The second line increments the pointer to the next integer, not the next byte. At this point, ip does not point to storage location 41; it points to storage location 44. Why? Because the compiler is smart enough to translate the following lines of C code differently:

```
char *cp=0;
int *ip=0;
cp++            =>      cp gets incremented by 1
ip++            =>      ip gets incremented by 4
```

Similarly, double-precision floating point numbers (data type *double*) occupy 8 bytes of storage each, therefore we have the following:

```
double *dp=0;
dp++            => dp gets incremented by 8
```

Back to the story-line. We have just incremented integer-pointer ip with the ip++ statement. At this point, ip does not point to storage location 41; it points to storage location 44. Then we increment it again with the following ip = ip + 1 statement; it now points to storage location 48 (which, by the way, is *outside* of array A). We now assign 0 to the storage location referenced by pointer ip:

```
*ip = 0;
```

This puts a zero into the first half of the double-precision floating-point number f and could cause quite a bug (see figure in the following page).

1.1.2   VARIABLES AND STACK ALLOCATION

Now you know what your C variables look like as they are stored in memory. One question to ask is "how does the machine operate on my variables?" In a load-store machine (one in which you must explicitly bring a value from memory into a register before you may operate on it), you cannot directly do things like this:

starting addr		< − − 32-bit quantity − − >			
52					
48	f:	0			
44		5551313			
40	A:	5551212			
36	cp:	18			
32		1	9	\0	?
28		i	z	e	
24		8	,		s
20		e	n		1
16	str:	s	t	\0	1
12	ip:	48			
8	c:				
4	b:				
0	a:				

```
a = b + c
```

where a, b, and c are memory locations.

You must instead move the values into registers, operate on them, and then write them back to memory (if their value has changed). The above code would translate to the following pseudo-assembly code:

```
lw    r1 <- MEM[ b ]
lw    r2 <- MEM[ c ]
add   r1 <- r1 + r2
sw    r1 -> MEM[ a ]
```

The first two instructions load the values of variables b and c into registers 1 and 2. The third instruction adds the two values together. Note that since memory has not changed, none of the variables have changed. The fourth instruction changes memory (by writing a value out to memory location a) and therefore changes the variable.

Important distinction: if you have not changed memory, you have not altered the variable.

This is because variables in C are located in memory, and, to operate on them, we must first bring them into registers; but this only brings in a copy of the variable, not the actual variable itself. In order to modify the actual variable, it requires a store instruction. An exception: C's *register variables* are variables that do not live in memory and are thus much faster, but not permanent.

Sidebar: an important distinction between character arrays and character pointers. Suppose we have the following:

```
char str1[] = "strlen 18, size 19";
char *str2  = "strlen 17, size 4";
```

This appears in memory as the following:

str2				"strlen 17, size 4"
	1	9	\0	?
	i	z	e	
	8	,		s
	e	n		1
str1 s	t	r		1

Thee first string, str1, is a character array, and space is allocated for it. When you declare an array, you must explicitly state how large it is, or implicitly state how large it is by giving it an initial value as is done here. Note that the size will never change, irrespective of the means by which the state is given.

The second string, str2, is a pointer to a character, and space is allocated for it, but the space allocated is enough to hold a pointer, not the entire array. The actual array contents (the string "strlen 17, size 4") is placed in an entirely different area of memory, and the pointer str2 merely references the string. Note that if you ever do something like this:

```
str2 = str1;
```

you will effectively lose the original contents of str2:

str2				"strlen 17, size 4"
	1	9	\0	?
	i	z	e	
	8	,		s
	e	n		1
str1 s	t	r		1

Str2 now points to the beginning of str1, and nothing points to the string "strlen 17, size 4" anymore—it is lost, effectively forever (the only way to find it again is to scan through memory doing pattern matches).

Functions in C work on the property of call-by-value, which means that if you send variables into a function, the values of those variables are copied into the local working-space of the function (its stack frame). When functions run, they are allocated temporary storage on the stack. This storage is effectively destroyed/consumed/garbage-collected as soon as the function stops running. Therefore, the following is an error:

```
char *
readline()
{
    char line[128];
    gets(line);
    return line;
}
```

It is an error because all of the variables belonging to a function and not declared *static* are considered off-limits after the function exits. However, none of the storage is actively reclaimed or deleted; it is just considered off-limits by convention, which means the values are still there until they are overwritten at some random point in the future—so your program may run just fine for a little while before it goes haywire. This makes debugging entertaining.

1.1.3   'RANDOM ACCESS' IS ANYTHING BUT

It is convenient to think about random access memory (both SRAM and DRAM) as what the name implies: storage technology that happily returns your data with the same latency every request, no matter what the address. This is not how reality works, however, as should become clear in the remainder of the lecture. When dealing with the cache as well as the with DRAM system, knowledge of your address patterns can help tremendously with application debugging and performance optimization, because different access patterns yield different latencies and different sustainable bandwidths.

Both caches and DRAMs have strong relationships with powers-of-two dimensions: turning memory addresses into resource identifiers (which bank, which row, which column, etc.) is much easier if the numbers and sizes of the resources are all powers of two, because that means one can simply grab a subset of the binary address bits to use as an identifier. Otherwise, one would have to do multiple passes of modulo arithmetic on every single cache or main-memory access. The power-of-two thing is important because software designers often like to use powers of two to define the sizes of their data structures, and this coincidence can ruin performance. As an example, consider the following code:

```
#define ARRAY_SIZE (256 * 1024)

int A[ARRAY_SIZE];
int B[ARRAY_SIZE];
int C[ARRAY_SIZE];

for (i=0; i<ARRAY_SIZE; i++)
     A[i] = B[i] + C[i];
```

Assuming that "int" maps to a 4-byte data quantity, each of these arrays is 1MB in size. Because the arrays are defined by the programmer one after another in the code, the compiler will arrange them contiguously in (virtual) memory. Therefore, if array A is located at the address 0x12340000, then array B will be found at location 0x1244000, and C will be found at 0x12540000.

This addressing is a critical issue when trying to access the items in those arrays. When the C code is executed on a general-purpose processor, the processor will first look inside its caches to see if the data is there. Assume the first-level cache is direct-mapped 64KB with 16-byte cache blocks; for the first iteration of the loop (accessing the first 4-byte words of each array), the hardware starts with a read to item B[0], takes address 0x12440000 and uses the address-subset 0x000 to index into the cache—this is the set of 12 bits between the '4' of the address and the trailing '0'. That's all that is used to determine where in the cache the controller will look for the data, and if the data is brought in from main memory, that is where that block will be placed in the cache. The hardware does the same for item C[0], so it uses the address 0x000 to find the datum. When the processor writes datum A[0] to the cache, address 0x000 is used as well. All three requests to item 0 of arrays A, B, and C all go to the same spot in the cache. Since the cache is direct-mapped, it cannot hold all of these requests, so all will miss the cache.

It gets worse.

16 bytes were fetched into the cache to satisfy each of these 4-byte requests, which means that, in a normal cache scenario, we should only have to bring in one data-fetch to satisfy three following requests. That is the main point of organizing caches into large blocks: when we move sequentially through memory, the following requests should "hit" in the cache, even if the first request misses. However, because each successive request to each different array overwrites the same cache block, when the processor gets around to the second iteration of the loop, in which it accesses datum B[1], that datum will not be found in the cache, and instead A[0] will be found there, which will cause the hardware to fetch B[1]. Furthermore, A[0] is recently-written data, and this data will have to be written out before B[1] can be fetched.

The end result: in the first four iterations of the loop, the block containing B[0], B[1], B[2], and B[3] is fetched into the cache four times. The block containing C[0], C[1], C[2], and C[3] is fetched into the cache four times. The block containing A[0], A[1], A[2], and A[3] is fetched into the cache four times and then written out to memory four times. In an ideal cache, there would have been three reads (or, in some caches, just two reads) and one write. In our case, there was four times that amount of memory traffic.

The problem gets a bit better out at the L2 (level 2) cache, because further out the caches are often set associative. This means that a cache is able to place several different cache blocks at the same location, not just one. So a four-way set-associative cache would not fall prey to the problems above, but a two-way cache would.

Once we move to the off-chip L3 cache, all addresses are certainly physical addresses, instead of virtual ones[1]. The operating system manages virtual-to-physical mappings at a 4K granularity, and for all intents and purposes in the steady state the mappings are effectively random [for more detail on virtual memory, see (Jacob and Mudge, 1998a),(1998b), and (1998c), and the Virtual Memory chapter in (Jacob et al., 2007)]. The compiler works in the virtual space, but (for general-purpose machines) at compile time it has no way of knowing the corresponding physical address of any given virtual address; in short, the compiler's notion of where things are in relationship to one another no longer holds true at the L3 cache. So, for instance, datum A[0] is almost certainly not located exactly 1MB away from datum B[0] in the physical space, just as datum A[1024] is almost certainly not located exactly 4KB away from A[0] in the physical space.

The end result is that this changes the rules significantly. Virtual memory, in the steady state (i.e., after the operating system has allocated, freed, and reallocated most of the pages in the memory system a few times, so that the free list is no longer ordered), effectively randomizes the location of code and data. While this undoes all of the compiler's careful work of ensuring non-conflicting placement of code and data [e.g., see (Jacob et al., 2007), Chapter 3], it also tends to smear out requests to the caches and DRAM system, so that you reduce the types of resource conflicts described above that happen at the L1 cache—in particular, at both the L3 cache level and the DRAM level (the DRAM's analogue to cache blocks is an open row of data, and these can be many kilobytes in size), the code snippet above would not experience the types of pathological problems seen at the L1 level.

However, plenty of other codes *do* cause problems. Many embedded systems as well as high-performance (supercomputer-class) installations use direct physical addressing or have an operating system that simply maps the physical address equal to the virtual address, which amounts to the same thing. Sometimes operating systems, for performance reasons, will try to maintain "page-coloring" schemes (Taylor et al., 1990; Kessler and Hill, 1992) that map virtual pages to only a matching subset of available physical pages, so that the virtual page number and the physical page number are equal to each other, modulo some chosen granularity.

More commonly, algorithms access data in ways that effectively convert what should be a sequential access pattern into a *de facto* strided access pattern. Consider the following code snippet:

```
struct polygon {
      float x, y, z;
      int value;
} pArray[MAX_POLYGONS];
```

---

[1] L2 caches in most processors today are still on-chip and, though they are typically physically indexed, can be virtually indexed.

```
for (i=0; i<MAX_POLYGONS; i++) {
        pArray[i].value = transform(pArray[i].value);
}
```

This walks sequentially through a dynamic list of data records—sequential access pattern, right? Effectively, no—the distance from each access to the next is actually a *stride*: a non-zero amount of space lies between each successively accessed datum. In this example, the code skips right over the x, y, and z coordinates and only accesses the values of each polygon. This means that, for every 16 bytes loaded into the processor, only 4 get used. We chose the struct organization for convenience of thinking about the problem and perhaps convenience of code-writing, but we're back to the pathological-behavior case. Better to do the following:

```
struct polygon {
        struct coordinates {
                float x, y, z;
        } coordinate[MAX_POLYGONS];
        int value[MAX_POLYGONS];
} Poly;
for (i=0; i<MAX_POLYGONS; i++) {
        Poly.value[i] = transform(Poly.value[i]);
}
```

This is a well-known transformation, called *turning an array of structs into a struct of arrays*. The name comes from the fact that, in the general case, e.g., if there are no groups of data items that are always accessed together, one might do something like the following:

```
struct polygon {
        float xArray[MAX_POLYGONS];
        float yArray[MAX_POLYGONS];
        float zArray[MAX_POLYGONS];
        int vArray[MAX_POLYGONS];
} Poly;
```

This would make sense if one were to perform operations on single coordinates instead of the x,y,z triplet of each polygon.

The notion works even for dynamic data structures such as lists and trees. These data structures facilitate rapid searches through large quantities of (changing) data, but merging the indices with the record data can cause significant inefficiencies. Consider the following code:

```
struct record {
        struct record *left;
        struct record *right;
        char key[KEYSIZE];
```

```
          /* ... many bytes, perhaps KB, of record data ... */
    } *rootptr = NULL;
    struct record *findrec(char *searchKey)
    {
        for (struct record *rp = rootptr; rp != NULL; ) {
            int result = strncmp(rp->key, searchKey, KEYSIZE);
            if (result == 0) {
                return rp;
            } else if (result < 0) {
                rp = rp->left;
            } else {   // result > 0
                rp = rp->right;
            }
        }
        /* if we fall out of loop, it's not in the tree */
        return (struct record *)NULL;
    }
```

Even though this might not seem amenable to optimization, it can be, especially for large databases. If the record-data size is on the order of a kilobyte (not unreasonable), then each record structure would occupy a kilobyte, and it only takes 1000 records to reach 1MB, which is significantly larger than an L1 cache. Any database larger than 1000 records will fail to fit in the L2 cache; anything larger than 10,000 records will fail to fit in even an 8MB L3 cache. However, what if we do the following to the data:

```
struct recData {
        /* ... many bytes, perhaps KB, of record data ... */
};
struct recIndex {
        struct recIndex *left;
        struct recIndex *right;
        char key[KEYSIZE];
        struct recData *datap;
} *rootptr = NULL;
```

If the key value is small, say 4 bytes, then the recIndex structure is only 16 bytes total. This means that one could fit the entire binary index tree for a 2000-record database into a 32KB level 1 cache. An 8MB L3 cache could hold the entire index structure for a half-million-record database. It scales beyond the cache system: this technique of partitioning index from data is common practice for the design of very large databases, in which the indices are designed to fit entirely within main memory, and the record data is held on disk.

Lastly, another way the strided-vs-sequential access issue manifests itself is in the implementation of nested loops. Writing code is deceptively simple: the coder is usually working very hard at the intellectual problem of making an algorithm that works correctly. What is not immediately obvious is that the structure of the code frequently implies an access pattern of the data. Consider the following two code snippets:

```
int pixelData[ HEIGHT * WIDTH ];
for (i=0; i<HEIGHT; i++) {
        for (j=0; j<WIDTH; j++) {
                compute( &pixelData[i * WIDTH + j] );
        }
}
for (j=0; j<WIDTH; j++) {
        for (i=0; i<HEIGHT; i++) {
                compute( &pixelData[i * WIDTH + j] );
        }
}
```

The only difference is that the loops have been exchanged. Assuming no hidden consequences from the compute() function, both code snippets should do exactly the same thing. And because both loops iterate one step at a time, it is tempting to think that both code snippets walk through memory sequentially, one integer at a time. However, while the first snippet walks through the pixelData array one integer at a time, the second snippet walks through the pixelData array touching one integer out of every WIDTH integers. For example, if the image size is 2000x3000 pixels, then the access pattern (in array indices) of the first code snippet is the following:

```
0, 1, 2, 3, ...
```

This is sequential. It is predictable, and it uses every byte fetched from the memory system. By contrast, the access pattern of the second code snippet is the following:

```
0, 3000, 6000, 9000, ... , 5994000, 5997000, 1, 3001, 6001, ...
```

It should be clear that, while predictable, this sequence is hardly sequential, and it only uses one integer out of every cache block fetched.

1.2   PERFORMANCE PERSPECTIVE

The three main problem areas of the DRAM system today are its latency, its capacity, and its power dissipation. Bandwidth has largely been addressed by increasing datarates at an aggressive pace from generation to generation (see the left side of Figure 1.1) and by ganging multiple channels together. That is not to say bandwidth is a solved problem, as many people and applications could easily make use of 10–100x the bandwidth they currently have. The main point is that, in the past decade, the

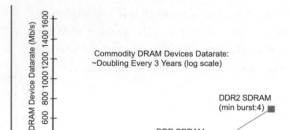
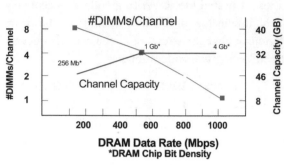

Figure 1.1: Trends showing datarate scaling over time (left), and channel capacity as a function of channel datarate (right).

primary issue addressed at the DRAM-system level has been bandwidth: while per-device capacity has grown, per-system capacity has remained relatively flat; while bandwidth-related overheads have reduced, latency-related overheads have remained constant; while processor power has hovered in the 100W range, DRAM-system power has increased, and in some cases can exceed the power dissipation of the processor.

Power will be left to the end of the chapter; one of the most visible problems in the DRAM system is its performance. As mentioned, bandwidth has been addressed to some extent, but main-memory latency, as expressed in processor clock cycles, has been increasing over time, i.e., getting worse. This was outlined famously by Wulf and McKee (1995) and termed "the memory wall." Each generation of DRAM succeeds in improving bandwidth-related overhead, but the latency-related overhead remains relatively constant (Cuppu et al., 1999): Figures 1.2(b) and 1.2(c) show execution time broken down into portions that are eliminated through higher bandwidth and those that are eliminated only through lower latency. When following succeeding generations of DRAM (e.g., when moving from Fast Page Mode [FPM] to EDO to SDRAM to DDR), one can see that each generation successfully reduces that component of execution time dependent on memory bandwidth compared to the previous generation—i.e., each DRAM generation gets better at providing band-width. However, the latency component (the light bars) remains roughly constant over time, for nearly all DRAM architectures.

Several decades ago, DRAM latencies were in the single-digit processor-cycle range; now they are in the hundreds of nanoseconds while coupled to processors that cycle several times per nanosecond and can process several instructions per cycle. A typical DRAM access is equivalent to roughly 1000 instructions processed by the CPU. Figure 1.3 shows access-latency distributions for two example benchmark programs (ART and AMMP, both in the SPEC suite): latencies are giving

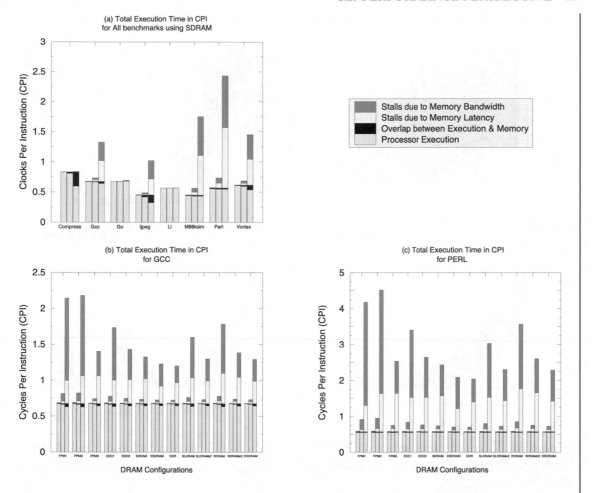

Figure 1.2: Total execution time including access time to the primary memory system. Figure (a) shows execution time in CPI for all benchmarks, using Synchronous DRAM. Figures (b) and (c) give total execution time in units of CPI for different DRAM types. The overhead is broken into processor time and memory time, with overlap between the two shown, and memory cycles are divided into those due to limited bandwidth and those due to latency.

on the x-axis, in nanoseconds, and number of instances is given on the y-axis. An interesting point to note is the degree to which clever scheduling can improve latency—for instance the dramatic improvement for ART by moving from a simple FIFO scheme to the Wang algorithm—but as the AMMP results show (in which the latencies of many requests are reduced, but at the expense of significantly increasing the latencies of other requests), such gains are very benchmark-dependent.

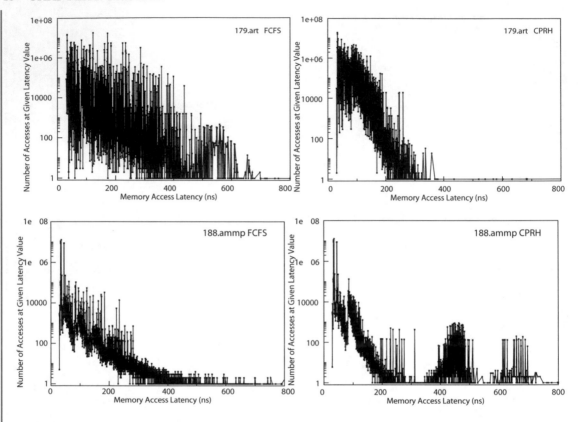

Figure 1.3: Range of access latencies for ART (one of the SPEC benchmarks), under two different scheduling heuristics: first-come, first-served (left) and Wang scheduling (CPRH, right). Figure taken from (Jacob et al., 2007).

Note that numerous techniques exist in the CPU to tolerate DRAM latency: these are mechanisms designed specifically to keep the CPU busy in the face of a cache miss, during which the processor would otherwise be stalled for many cycles waiting for the DRAM system. Examples include lockup-free caches, out-of-order scheduling, multi-threading, and prefetching. As should be intuitively obvious, these enable the CPU to get work done while waiting for data to return from main memory; they tend to provide this benefit at the cost of increased power dissipation and a need for additional memory bandwidth.

Another issue is the capacity problem. Over the years, as signaling rates (and therefore channel datarates/bandwidths) have increased, the maximum number of DIMMs per channel has decreased—a necessary sacrifice to ensure signal integrity on the increasingly high-speed data bus. The right side of Figure 1.1 illustrates these trends. Total capacity in the channel has been relatively flat for a relatively long period of time: working sets have increased in size; DRAM device density and thus capacity has increased; per-DIMM capacity has increased; but total channel capacity has *not* increased. Unfortunately, applications have not held still; they have increased in size and scope, and with that increase has come an increased need for memory storage.

Capacity is a very real, immediate problem. Not enough DRAM in a system limits the performance available from that system—in particular, consider a server system in an enterprise-computing scenario: workload is added to each server (often in the form of additional copies of the same programs) until each machine begins to page to disk. Once this happens, the machine's load can no longer be increased (and might need to be decreased slightly), because performance becomes unacceptable if the disk is any significant part of the equation. The capacity and speed of the memory system directly determine the degree to which a system pages.

The secondary effect from reduced DRAM capacity is that, by forcing administrators to use additional servers when they wish to increase load, the total number of servers executing must increase, thereby increasing total power dissipation. Bottom line: fit more DRAM into a chassis, and you can reduce your datacenter's electricity bill.

1.3   MEMORY-SYSTEM ORGANIZATION AND OPERATION

This section gives the reader an intuitive feel for the way that the DRAM system behaves, which is arguably more important for a high-level CS reader than a detailed description of its inner workings. For those who want enough detail to build their own memory system, other reports provide more comprehensive technical info (Cuppu et al., 1999; Cuppu and Jacob, 2001; Cuppu et al., 2001; Jacob et al., 2007). The main point to take away in this section is that the memory system's behavior is driven by request scheduling. This is important because scheduling is commonly known to be an NP-complete problem, which implies two things: first, it is a challenging problem, and second, it is hard to do provably well. Thus, it should come as no surprise that building a memory system that performs well is non-trivial, and designing one's hardware and/or software to work well with the memory system is just as non-trivial.

Figure 1.4 shows a stylized general purpose memory hierarchy: at the top is a CPU with its cache subsystem, below that is the DRAM subsystem, and below that is the disk subsystem. The disk provides permanent store at very low cost per bit; the DRAM provides operating store, meaning random access and small granularity of access (bytes or tens of bytes per access instead of hundreds or thousands of bytes per access), at a higher cost per bit than disk; and the cache provides high bandwidth access to select data and instructions, at a cost per bit much higher than that of disk or DRAM. Consequently, the hierarchy is usually represented with a pyramid shape, indicating that there is typically several orders of magnitude or so difference in capacity between each level: at the consumer level, caches are on the order of megabytes in size; DRAM systems are on the order of gigabytes; and even desktop disk systems are rapidly approaching the terabyte mark (as a datapoint, I have a terabyte RAID drive in my basement that serves the household).

Figure 1.4: Stylized memory hierarchy.

Figure 1.5 shows a detail of the DRAM system, illustrating the various levels of organization in a modern system. A system is comprised of potentially many independent DIMMs. Each DIMM may contain one or more independent ranks. Each rank is a set of DRAM devices that operate in unison, and internally each of these DRAM devices implements one or more independent banks. Finally, each bank is comprised of slaved memory arrays, where the number of arrays is equal to the data width of the DRAM part (i.e., a x4 part has four slaved arrays per bank). Having concurrency at the rank and bank level provides bandwidth through the ability to pipeline requests. Having multiple DRAMs acting in unison at the rank level, and multiple arrays acting in unison at the bank level, provides bandwidth in the form of parallel access.

Figure 1.6 shows the detail of a DRAM device, corresponding of one or more DRAM arrays in Figure 1.5. Data is stored as charge or lack of charge on a capacitor; each DRAM device has billions of very small, but very powerful capacitors. Data is extracted from this array by first precharging the array (raising the potential on each bit line to a voltage halfway between a 0 and a 1), then selecting an entire row's worth of data (called a *row activation*) by pulling one word line high (thereby turning

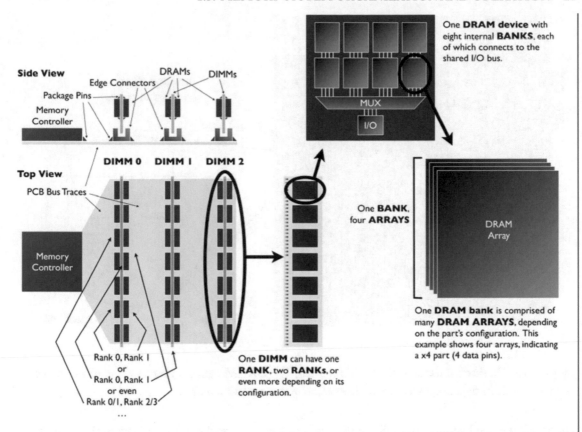

Figure 1.5: DIMMs, ranks, banks, and arrays. A system has potentially many DIMMs, each of which may contain one or more ranks. Each rank is a set of ganged DRAM devices, each of which has potentially many banks. Each bank has potentially many constituent arrays, depending on the parts data width.

on all of those pass transistors), then *sensing* the data on those bit lines and recovering the full 0/1 voltage values, and finally taking a subset of the sensed data and driving it out the I/O pins to the memory controller on the other side of the bus (called a *column read*).

Figure 1.7 shows the rough time delays relative to each other: a precharge places the bank in a voltage state that is suitable for activation, an activation reads an entire row of data bits (typically 8000 bits per DRAM device) into the sense amplifiers, and a column read or write transfers data to or from the memory controller. Usually precharge can be hidden to an extent by doing it as early as possible, even before the next request is known. A *close-page* operation is an explicit or implicit precharge that makes the bank ready for the next request (it is part of a following request but is usually done as soon as possible, even before the next request is known). A precharge operation is

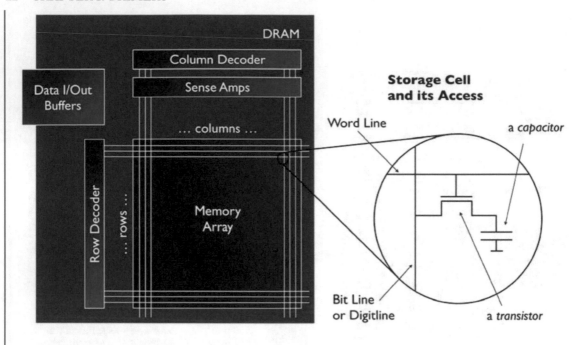

Figure 1.6: Basic organization of DRAM internals. The DRAM memory array is a grid of storage cells, where one bit of data is stored at each intersection of a row and a column.

optional in some instances wherein multiple requests want to read or write data from the same large row, in which case it may make no sense to repeatedly open and close that row.

Note that each of these operations (precharge, activate, column access) takes anywhere from a few nanoseconds to several tens of nanoseconds. Compare this to the handful of nanoseconds that it takes to transfer an entire burst of data bits over the bus, and it is easy to see why DRAM is considered a high-overhead technology: even for long data bursts (DDR3 is fixed at 8 beats of data for every read/write access), the cost of transferring the data is only a quarter of the total amount of time spent on the access. It should come as no surprise, then, that it takes *significant* effort to amortize the high access costs so that they do not become part of the visible access latency.

Figure 1.8 gives a stylized illustration of what is meant by "significant effort." As many people are aware, a typical system today supports many high-performance concepts that have come out of the supercomputer community over the last few decades, namely

- Caching

- Pipelining

- Concurrency

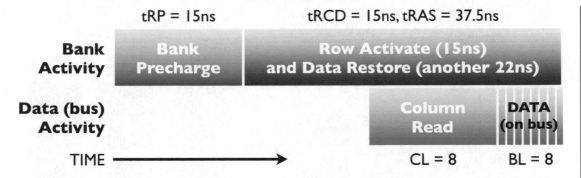

Figure 1.7: The costs of accessing DRAM. Before a DRAM cell can be read or written, the device must first be placed into a state that allows read/write access to the cells. This is a two step process: the **precharge** command initializes the banks sense amplifiers (sets the voltage on the bitlines appropriately), and the **activate** command reads the row data from the specified storage capacitors into the sense amplifiers. Once the sense amplifiers have read the data, it can be read/written by the memory controller. Precharge and activate together take several tens of nanoseconds; reading or writing takes roughly a nanosecond. Timing values taken from a Micron 2Gb DDR3-1066 part.

- Reordering

But, of course, we are not talking about the CPU here (which everyone knows supports these things); we are talking about the memory system. Every modern memory system today has a significant degree of caching, pipelining, concurrency, and reordering going on in it. The hallmarks of a high-performance DRAM system are numerous resources all acting concurrently with a considerable amount of scheduling, queueing, reordering, and management heuristics controlling the movement, timing, and granularity of commands and data. Among other things, having multiple banks (and multiple ranks of banks) within a channel allows different banks to handle different commands at different times—at any given moment, several different banks can be precharging, several others can be activating and reading out rows of data into their sense amplifiers, and yet another can be reading or writing data on the bus. Due to the nature of a bus, only one bank can be reading or writing column data on the bus at a given moment, but a system can provide additional concurrency in the form of multiple independent channels that *can* handle simultaneous bus reads and writes.

Looking at the figure, at the top left, a processor is placing a request on the bus that connects it to the memory controller (which could be on-chip or off-chip—the behavior is the same). Note that the format of the request is something that the CPU can understand: it is a read or write request, and it is to a memory block in the CPU's notion of the physical memory address space (example assumes that the CPU has already translated the virtual address to physical space, the typical scenario).

This request is placed on the bus interface unit's queue along with other similar requests; note that data to be written accompanies write requests, as is indicated in the figure. The memory

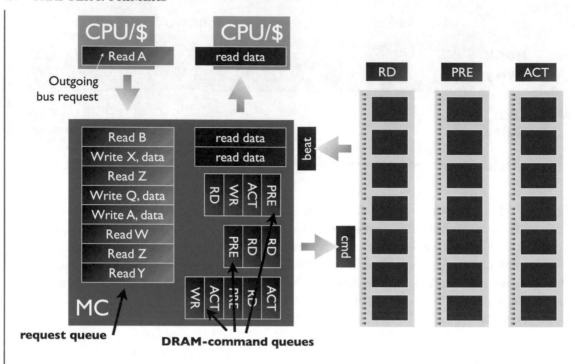

Figure 1.8: Stylized DRAM system.

controller pulls requests out of this queue and places them into DRAM-request queues. The order in which the requests are pulled out of the incoming request queue is determined by the memory controller's request-scheduling policy: for instance, this could be in order (a FIFO queue)—or it could be out of order, prioritized by common addresses, bank conflicts, request age, read/write ordering priority (reads typically take precedence over writes), bus utilization, or any number of other factors.

As mentioned, requests are pulled out of the incoming request queue, dictated by the memory controller's request-scheduling policy; they are transformed into DRAM command sets and placed into one or more queues awaiting availability of the DRAM devices. There is one logical DRAM-command queue for every independent DRAM channel supported; each queue corresponds to an independent channel managed by an independent memory controller. Channels that are ganged together for bandwidth purposes typically share a controller and thus a queue. The physical organization of the DRAM-command queue is implementation dependent; existing implementations include monolithic queues but are more usually divided into multiple queues corresponding to different physical resources; this division aids the command-scheduling heuristic. Typical organizations include per-rank queues and per-bank queues—per-bank in both flavors: one where all "bank i" requests go to the same queue, regardless of rank, and one in which every unique bank gets its own unique queue. Note that a per-bank queue is possible in modern systems that significantly limit the number of ranks per channel (often to two); were a channel limited to a much larger number of ranks, the number of queues in the memory controller could become unwieldy.

The DRAM-command queues hold requests in a form that a DRAM device would understand: the physical addresses in the incoming request queue are translated to channel number, rank number, row number, column number. Commands placed in the DRAM-command queues include PRE (precharge), ACT (activate, or row access), RD (column read), WR (column write), or reads and writes with autoprecharge (closing out the row).

Commands, once generated by the controller and placed into these DRAM-command queues, wait until their channels are ready; like the request queue, the command queues may be in-order (FIFO), or they may be scheduled. A simple but powerful example implementation uses per-bank queues, maintains the queues in FIFO order, and schedules commands on the bus by selecting between queues, choosing the command at the head of a queue that can fire the soonest, uses the least resources, finishes the soonest, maximizes concurrency, or adheres to any number of other heuristic metrics of efficiency.

It should be clear to see that, just within the memory controller, there is significant queueing and scheduling (and potentially reordering) of requests. When DRAM commands are sent to the devices on a particular channel, there is typically no more queueing or scheduling, and any read data returns via the data bus to the memory controller in a predictable order and with a predictable latency.

Data is sent to the DRAM devices and returns to the controller via the data bus in multiple *beats*, as is shown in the figure: though a request is notionally atomic, it usually requires multiple clock cycles to transfer an entire burst between DRAM devices and memory controller. Typical burst lengths are 4 and 8. Given a typical general-purpose bus width of 64 bits (8 bytes), this represents 32 or 64 bytes per request, corresponding to typical cache-block sizes.

Data returning to the memory controller is queued up for transmission to the requesting CPU. The buffering is implemented to allow the two bandwidths to differ: i.e., a memory controller will work perfectly well even if the bandwidth on the DRAM side is not exactly equal to the bandwidth on the CPU side. In the case of split-transaction or other forms of reordering busses, transaction IDs can be necessary to identify returning requests.

There is significant queueing and scheduling within the memory system; in addition, the various queues within the controller and the sense amplifiers within the DRAM devices all provide significant degrees of caching, and the simultaneous operation of all these mechanisms provides for a significant degree of pipelining from beginning to end. Indeed, without significant pipelining, the full DRAM latency could be exposed on nearly every request, which in real systems is not generally the case.

1.4   STATE OF THE (DRAM) UNION

Today one can buy a 2GB memory stick i.e., "DIMM" or "memory module" … a circuit board with DRAM chips on it) for about $20. Realize that this is a commodity part: it works in desktops and low-end servers; slightly different memory sticks that cost a few dollars more and use smaller form factors (called SO-DIMMs: "small outline" DIMMs) go into laptops. These commodity desktops, laptops, and low-end servers have memory channels that can accept up to two DIMMs (some laptops, for space considerations, will only accept one). So: by purchasing 2GB DIMMs, one can have a maximum of 4GB in a DRAM channel. By paying a little more for 4GB sticks, one can put 8GB in one's system. By paying even more to attach two channels to the processor, like many server systems do, you can double this and put 16GB into your system. Simple enough.

The same commodity DRAM chips in those 2GB and 4GB commodity DIMMs are used in non-commodity DIMMs as well. High-end servers have extra engineering in their memory systems (both active and passive electronics) that support a maximum of not two but *four* DIMMs per channel. They also use non-standard form factors and put more DRAM chips onto each module, generally increasing DIMM capacity by 4–8x over commodity systems. So, were one to purchase 16GB DIMMs, the memory system could have 64GB total, not just 8GB, per channel. Factor of eight. How much do these non-commodity DIMMs cost for a factor of eight increase in capacity? Roughly $2000. Two orders of magnitude.

Granted, this represents a minority share of the server-memory market—itself a minority share of the general DRAM market—but, even so, why is *anyone* willing to pay such a price premium? As mentioned, applications over time have increased in size and scope, and with that increase has come an increased need for memory storage. There are numerous applications out there that absolutely require increased capacity, which leads to the market supporting those price premiums.

A few years back, Intel sought to solve the capacity problem by developing a commodity DIMM and system architecture that could support at least an order of magnitude more memory at the system level. The *Fully Buffered DIMM (FB-DIMM)* was released as a JEDEC standard (JEDEC, 2007). Its increase in capacity came at a relatively significant price in total power dissipation: whereas DRAM systems were previously considered heat sinks in terms of system design (i.e., as far as the people responsible for power dissipation and cooling of components within the enclosure were concerned), a fully populated FB-DIMM system can easily exceed the power and cooling requirements of the CPU and disk.

There is still a holy grail in terms of total system capacity within the previous power and thermal envelope; several Silicon Valley start-ups are rushing to meet that need. Ignoring cost, there are four primary metrics that are important:

- Bandwidth

- Latency

- Power

- Capacity

Achieving **bandwidth** is easy if you are willing to give up low latency and add more power: for instance, go faster and/or wider. Faster busses *require* longer latencies (in bus cycles, not necessarily wall-clock time), and both solutions clearly dissipate more power. In such a scenario, **capacity** can come along for free—for instance, FB-DIMM is such a solution.

Achieving low **latency** is easy if you are willing to pay more money and more power—for example, one could build an entire memory system out of RLDRAM, which has low latency but is more expensive than vanilla DDR in both dollars and Watts. At the extreme end, you can replace DRAM with SRAM and be done with it, achieving *really* low latency.

If you want more **capacity**, you have to pay more power (to a first order, more devices = more power).

If you want low **power**, you have to sacrifice capacity, bandwidth, and latency—for instance, something like PSRAM is a potential solution. In addition to those sacrifices, it will also be expensive per bit.

So there you have it. The one thing you have to sacrifice for each of the desired metrics, the one thing we can't get away from is power. The future of DRAM systems has to include more managed power states, for instance smarter devices that can manage their own power states, smarter DIMMs that can manage their local state, and smarter controllers that manage power at a system level. Hur and Lin's recent work (Hur and Lin, 2008) is an extremely good step, unfortunately their specific algorithm won't work in a DDR2 memory system[2]. Following the "smarter" concept to its logical conclusion produces non-slave memory devices that can manage their own power states and that interact with the memory controller on a peer-to-peer basis.

Just as the microprocessor industry was forced to back down from the high frequency CPU path to chase power-efficient microarchitecture, so, too, the DRAM and memory-system industries need to chase power-efficient designs. It may even be more important for memory, because, as you put more cores per CPU socket (the current trend), you need to increase the per-CPU-socket memory footprint, otherwise memory-per-core and thus performance-per-core will drop.

Power is a big problem in every segment of the market space. It is a show stopper in some, and a huge annoyance in others. Alleviating the power problem for more "breathing room" simplifies the search for more bandwidth, lower latency, and higher capacity.

---

[2] You cannot quite power manage DDR2 SDRAM devices in a multi-DIMM system in the way Hur & Lin want to; the termination functionality relies on the DLL for accurate timing. When you do power management, the DLL gets turned off, and you don't get very accurate timing on ODT control. Thus, you have to wait 3–4 cycles to make sure you get a nicely terminated system—per column access. However, if you do that, you end up losing two-thirds of your bandwidth just trying to manage your power.

CHAPTER 2

It Must Be Modeled Accurately

(you can't fake it; do it right or don't bother doing it at all)

In modern medium- to high-performance computer systems, the memory system is relatively important in determining the performance of the overall system—more so than, say, the CPU microarchitecture. This is by no means a new phenomenon; the computer-design community has known it since at least the mid-1990s:

> In a recent database benchmark study using TPC-C, both 200-MHz Pentium Pro and 400-MHz 21164 Alpha systems were measured at 4.2–4.5 CPU cycles per instruction retired. In other words, three out of every four CPU cycles retired zero instructions: most were spent waiting for memory. Processor speed has seriously outstripped memory speed.
>
> —Richard Sites. "It's the memory, stupid!" (Sites, 1996)

Numerous reports throughout the 1990s and early 2000s demonstrated a similar phenomenon: processors that should, in theory, sustain 0.5 CPI were instead running at closer to 5 CPI once memory-system details were taken into account. Note that most of these reports were machine measurements from industry and not just academic simulation-based studies, a point that is quite significant (the implication is that the measurements were accurate, and the theoretical studies were not). It turns out that, in modern systems, roughly 90% of a computer system's time is spent not computing but waiting on the memory system: the caches, the DRAMs, the disks.

This fact surprised many—in particular, the many who failed to use accurate models of the memory system, as it was this use of simplistic models of memory performance that predicted the numerous sub-1.0 CPI figures to begin with.

If we, as a community, expect to obtain accurate results (i.e., no more getting blind-sided by implementations that fail to match predicted performance by an order of magnitude), then we would do well to observe the following rule of thumb:

The model's accuracy must grow with the complexity of the system to be modeled.

… for instance, if the system to be modeled is simple, a simple model suffices; beyond that, results from simple simulators diverge *quickly* from reality. Complex systems need correspondingly detailed models for accurate, detailed, specific (as opposed to aggregate) measurements.

What this means is straightforward and should be intuitively obvious. The performance of those systems that use simple uniprocessors (e.g., in-order, blocking pipelines) can be modeled accurately with extremely simple models of the memory system—for instance, using a constant memory access time to satisfy a last-level cache miss. However, as complexity is added to the system in the form of lock-up free caches, superscalar and out-of-order execution, multiple threads and/or cores, prefetching, branch prediction, code & data speculation, etc., then the simple memory model is found to misrepresent reality by anywhere from 10% to a factor of ten, in both directions (over- and under-prediction).

The bottom line: unless you are working with simple systems, you can no longer fake the memory system in your performance simulations (i.e., you cannot use a simplistic model), because doing so is guaranteed to produce *garbage in, garbage out* results.

An anecdote. Numerous academic studies have predicted that using prefetching in server applications should be beneficial, but this is contradicted by industry practice, which is to ***disable*** the prefetching units of microprocessors when they are used in server settings (as opposed to desk- top/laptop settings). That is significant, so I will repeat it:

> Despite the fact that common wisdom, backed up by numerous academic studies, holds that prefetching should benefit server applications, prefetching is routinely *turned off* in server machines, because leaving it on *degrades* performance. Off good. On bad.

A recent study shows exactly why this should be the case and why there should be such confusion within the community (Srinivasan, 2007). When these prefetching studies are repeated using an *accurate* model of the memory system, there is no longer a disconnect—the results mirror what industry has known for years: server applications compete with their prefetching mechanisms for the available memory bandwidth, and so one must completely re-think how prefetching is done in such environments.

In retrospect, it should not be at all surprising that blindly transplanting into a heavily loaded multicore environment a prefetch scheme that works well in a lightly loaded uniprocessor environ- ment might represent an approach that is not immediately successful. One obvious conclusion from reading the study is that all prefetching schemes evaluated within the context of a simple memory model are now suspect; they should all be re-evaluated within the context of a real memory system. This, of course, extends far beyond prefetching and indeed applies to every single high-performance gadget proposed in the last decade: all results are suspect until evaluated in the light of a realistic memory system.

The rest of this chapter is a condensed form of that study (Srinivasan, 2007).

2.1 SOME CONTEXT

Figure 2.1 shows the bandwidth vs. latency response curve in a system with a maximum sustained bandwidth of 10 GB/sec. Maximum sustained bandwidth is the maximum bandwidth observed in the system and is different from the theoretical maximum. Maximum sustained bandwidth is often 65% to 75% of the theoretical maximum for server workloads, and depends on various factors such as read-write ratio, paging policies, address mapping, etc. This type of curve shows, for an observed aggregate bandwidth usage over time, the corresponding average latency per request, during that same time. Effectively, this shows the cost, in access time, of using the available bandwidth, up to the

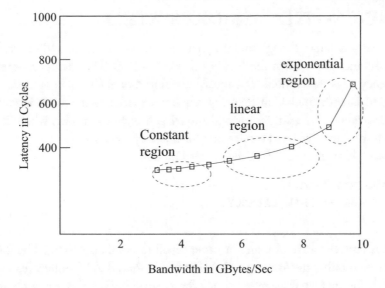

Figure 2.1: Bandwidth vs. latency curve. The memory bandwidth vs. average memory latency for a system with a maximum sustained bandwidth of 10 GB/sec. is shown. Memory latency increases exponentially as the amount of bandwidth used approaches the maximum sustainable bandwidth.

maximum sustainable amount (which, again, is substantially lower than the maximum theoretical bandwidth). The bandwidth-latency curve consists of three distinct regions.

Constant region: The latency response is fairly constant for the first 40% of the sustained bandwidth. In this region the average memory latency equals the idle latency in the system. The system performance is not limited by the memory bandwidth in this zone, either due to applications being non-memory bound or due to excess bandwidth availability.

Linear region: In this region, the latency response increases almost linearly with the bandwidth demand of the system. This region is usually between 40% to 80 % of the sustained maximum. The average memory latency starts to increase due to contention overhead introduced into the system by numerous memory requests. The performance degradation of the system starts in this zone, and the system is claimed to be fairly memory bound.

Exponential region: This is the last region of the bandwidth-latency curve. This region exists between 80%–100% of the sustained maximum. In this zone the memory latency can be many times the idle latency of the DRAM system. Applications operating in this region are completely memory bound, and their performance is limited by the available memory bandwidth.

The figure illustrates the need for a system to operate in the *constant region* or at worst the *linear region*. The need for a better memory system becomes even more important as we go towards aggressive multi-core systems, and simple models are increasingly inaccurate in these scenarios.

2.2   MODELING THE MEMORY SYSTEM

Most modern CMP simulators, though they might advertise detailed cache and interconnect models, assume a simplistic memory-system model (Martin et al., 2005). The memory system is assumed to be a fixed latency or is represented by a simple queuing model (Emma et al., 2005; Zhao et al., 2007). In the fixed latency model, all memory requests experience the same latency irrespective of bandwidth constraints. A slightly improved model is a queuing model, which has bandwidth constraints with a specific arrival and service rate for memory requests.

The following is an example of a fixed-latency model:

```
if (cache_miss(addr)) {
        cycles += DRAM_LATENCY;
}
```

This code is representative of many in-order pipeline simulators today. The `DRAM_LATENCY` constant can represent either the lowest possible latency of the DRAM system (e.g., an open-page DRAM bank hit), the measured average latency for a particular benchmark, a conservative value larger than the measured average, or anything in this neighborhood, but the main points are that it is a constant and that it is assessed at the time of the last-level cache miss. These two points have significant consequences: the first point leads to the implication that no queueing, scheduling, or reordering of memory requests will happen once the request leaves the processor; the second point implies that either the processor stalls on cache misses, or memory bandwidth is unlimited. The code above suggests that the processor stalls on all cache misses, but consider the following re-write of the code, handling data accesses:

```
if (cache_miss(instr.loadstore_addr)) {
        instr.loadstore_available = now() + DRAM_LATENCY;
}
```

This code simply uses the memory-system latency to tag the target register with the time at which the data becomes available (ultimately). If the processor model implements non-blocking caches, then the simulator implicitly allows multiple outstanding memory-system requests to be in flight, all potentially using the same physical resources such as the same DRAM bank and the same memory bus, all at the same time. This practice can be nothing short of disastrous, and it should be noted that this code is used in most out-of-order simulators today. Why this relates to prefetching should be obvious in the following rewrite:

```
if (cache_miss(instr.loadstore_addr)) {
        instr.loadstore_available = now() + DRAM_LATENCY;


        for (i=0; i<PREFETCH_DEPTH; i++) {
                prefetchQ_t *entryp = allocate_prefetch_entry();
```

```
    entryp->loadstore_addr =
        predict_ith_prefetch_addr(instr.loadstore_addr, i);
    entryp->loadstore_available = now() + DRAM_LATENCY;
    }
}
```

This code (or code that is functionally similar) has been used to perform nearly all prefetching studies to date. The problem is that this model immediately *over*-saturates the memory bus with prefetch requests, without paying for them, and so it should not be the least bit surprising that prefetch studies using code like this will return results that make their prefetch algorithms look good. Intuitively, *any* reasonable algorithm should look good if you can prefetch as many things as you want on a cache miss, into a prefetch buffer, without having to pay the bandwidth cost.

As it turns out, the only way that we have been able to reproduce performance results from many, many different prefetching studies is by using code similar to that last example.

The bandwidth issue can be addressed (i.e., bandwidth consumption can be limited to a realistic value) by using a more accurate model that keeps track of bus utilization, to ensure that a set bandwidth threshold is never exceeded. Intel research calls this a "queueing model" (Zhao et al., 2007) and relates its behavior to Poisson M/M/1 models (Kleinrock, 1975). We will simplify that discussion and just deal with some example code:

```
if (cache_miss(instr.loadstore_addr)) {
  // cheap & dirty solution for purposes  of illustration
  while (current_bandwidth()  >  SUSTAINABLE_BANDWIDTH) {
      wait();
  }
  // find when bus is available for a length of time (burst length),
  // given a soonest poss. starting time (min_latency, as follows:
  //    MIN_LATENCY + BURST_LENGTH = DRAM_LATENCY).
  // func returns whatever delay is required, on top of min_latency
  // (note: it could be zero, indicating no contention)
  delay = find_bus_availability(now() + MIN_LATENCY, BURST_LENGTH);
  // grab the bus from a starting time, lasting a duration
  reserve_bus(now() + MIN_LATENCY + delay, BURST_LENGTH);
  instr.loadstore_available = now() + DRAM_LATENCY + delay;
}
```

This code behaves much like the previous snippets, except that it ensures no more than one request is on the bus at a time. It implies in-order servicing of requests, and it guarantees that the number of requests during a given window do not exceed some max threshold which could be less than 100% (for example, realistic memory systems usually saturate at 70–80% of their maximum theoretical bandwidth).

Prefetching in such a model, to maintain the bandwidth limitations, would be implemented in a manner such as the following:

```
if (cache_miss(instr.loadstore_addr)) {

    while (current_bandwidth() > SUSTAINABLE_BANDWIDTH) {
        wait();
    }
    delay = find_bus_availability(now() + MIN_LATENCY, BURST_LENGTH);
    reserve_bus(now() + MIN_LATENCY + delay, BURST_LENGTH);
    instr.loadstore_available = now() + DRAM_LATENCY + delay;

    for (i=0; i<PREFETCH_DEPTH; i++) {
        prefetchQ_t *entryp;

        while (current_bandwidth() > SUSTAINABLE_BANDWIDTH) {
            wait();
        }
        entryp = allocate_prefetch_entry();
        entryp->loadstore_addr =
            predict_ith_prefetch_addr(instr.loadstore_addr, i);
        delay = find_bus_availability (now() + MIN_LATENCY,
                                       BURST_LENGTH);
        reserve_bus(now() + MIN_LATENCY + delay, BURST_LENGTH);
        entryp->loadstore_available = now() + DRAM_LATENCY + delay;
    }
}
```

Like the fixed-latency model earlier, the `DRAM_LATENCY` constant can represent either the lowest possible latency of the DRAM system, the measured average latency for a particular benchmark, a conservative value larger than the measured average, or anything in this neighborhood, but the main point is that it is a constant, implying that there is no reordering or coalescing of requests once they leave the processor. Though this last model is much more accurate than the fixed-latency model earlier, it still does not represent the true behavior of the memory system (it ignores, for instance, the state of the DRAM devices), and this difference can be significant.

2.3   COMPARING THE MODELS

Here we examine the impact of using fixed-latency and queuing models for CPU and system design as the complexity of the CPU is increased, represented primarily by the number of cores on-chip but also including advanced mechanisms such as prefetch. The bottom line is that as complexity is increased, the performance difference between these models and the cycle-accurate simulator

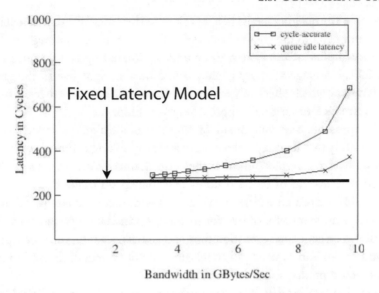

Figure 2.2: Fixed-latency model and queuing model vs. cycle-accurate model. This graph compares the memory latency predictions of a cycle accurate model with a less realistic models, for various bandwidths. The x-axis represents the various sustained bandwidths of the system, and the y-axis denotes the average memory latency corresponding to it. The queuing model assumes a Poisson arrival and service rate. The idle latency model is represented by the solid straight line for various bandwidths.

increases. Thus, as you add mechanisms such as multi-core, prefetching, lock-up free caches, out-of-order execution, branch prediction, and other high-performance mechanisms that tax the memory system, your performance predictions will get increasingly inaccurate if you use a simple memory model.

The baseline memory controller is a detailed cycle-accurate model that supports DDR and FBD protocols (Wang et al., 2005). The model supports various scheduling algorithms such as read first, write first, adaptive etc. The scheduling algorithm used in this study is an adaptive scheduling algorithm that gives priority to read requests over write requests as long as the number of outstanding writes is below a threshold, set at 2/3rd of the write queue size. The model also provides the flexibility to vary the address mapping policies, number of ranks, DIMMs etc. in the system. The other models (fixed latency and queueing models) are similar to the code presented earlier.

Figure 2.2 shows the difference in average memory latency of a system while using a cycle accurate model vs. a queuing model (all the other components in the system such as cache and interconnects being the same), as well as a fixed-latency model. Both cycle-accurate and queueing models were simulated for different system bandwidth requirements with a maximum of 10 GB/sec. and the same idle latency.

We observe that the queuing model behaves close to the cycle-accurate model in the constant region, but it does not capture the contention overhead accurately at other regions. This results in the average memory latency of the system to be underestimated by 2x in a queuing model. In a fixed latency model the average memory latency would be a straight line in the graph for all the bandwidth requirements, which effectively makes all bandwidth-intensive mechanisms appear free, and thus their performance-to-cost ratio approaches the infinite.

This graph represents one component of why the models fail to predict system behavior accurately. Next, we study how inaccurate those models are by looking at the execution time predicted by the models. In general, our results show that the performance difference between the models can be 15% in a multi-core system of in-order cores that lockup on cache miss, no prefetching, no branch prediction. This performance difference increases with increased bandwidth requirements of the system and can go up to a factor of two for memory optimization studies such as prefetching. This discrepancy between simulated execution time and actual execution time (as represented by the cycle-accurate model) can lead to wrongful conclusions about certain optimization techniques and can result in substandard products.

First, we take a look at the difference between the minimum latency in a DRAM system and the average latency in a DRAM system. The minimum latency is system specific, and the average latency is both system and application specific. Figure 2.3 illustrates, using SAP and SJBB benchmarks as examples. Figure 2.3(a) shows latency as a cumulative result; Figure 2.3(b) shows latency as a probability density graph. Both graphs use a single DDR800 channel (6.4 GB/s) as a baseline memory system, coupled to a 4GHz processor. See (Srinivasan, 2007) for details of the experimental setup, including descriptions of the benchmarks. Overlaid on the observed latency curves are vertical lines indicating the idle latency of the system (corresponding to the minimum DRAM-system turnaround time, as measured in 250ps processor cycles) and the application's measured average latency (corresponding to the average of the observed realistic DRAM-system turnaround times [measured via cycle-accurate simulation], also given in 250ps cycles). A true memory system, with complex queueing and scheduling mechanisms, will see a large distribution of latencies. Simple models will not reproduce this behavior accurately.

These two constant access times, the idle/minimum latency and the average latency, are each used in the fixed-latency and queuing models to produce four difference variations, all of which are compared against the cycle accurate model. They are labeled in the following legends as Fixed/Idle (FILM), Fixed/Average (FALM), Queuing/Idle (QILM), and Queuing/Average (QALM) Latency Models. In the following graphs, the performance of the cycle-accurate model is given by the horizontal lines, and the results of the other simulators are all shown relative to the cycle-accurate model.

As Figure 2.4 shows, the degree of inaccuracy introduced by the memory model increases with the number of cores. It is true for the simplistic fixed-latency model, and it is true even for the queueing model, which takes into account the first order of effects like the impact of bandwidth on memory latency. Figure 2.4 shows the performance (as IPC, instructions per cycle) of various

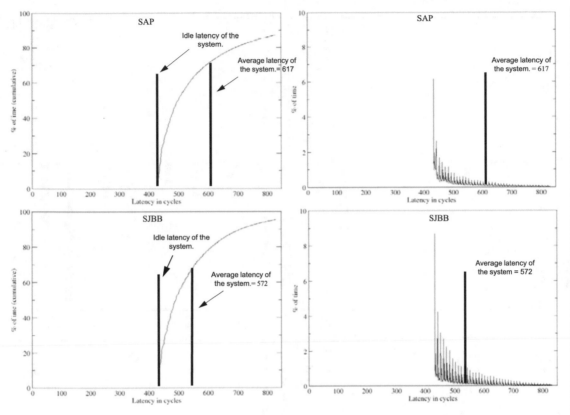

Figure 2.3: Memory latency response for DDR-800. The figure shows the memory latency distribution, idle and the mean latency for SAP and SJBB with 8 cores. The first solid line indicates the idle latency of the system: the minimum round trip time for any memory request. The second solid line is the average latency of the system. This latency is the mean of all memory-request latencies for the entire simulation run. Graphs on left (a) are cumulative; graphs on right (b) are probability density. The latency distribution is concentrated closer to the idle latency and tapers off gradually. Most of the latencies lie to the left of average latency, i.e. they are less than the mean latency of the system.

memory models normalized to a accurate memory controller (AMC) for SAP, SJBB, TPCC and SPECW. The x-axis represent the various cores and y-axis represent the IPC values of various models normalized to the cycle-accurate memory controller (AMC). In the figures, a negative values indicates that the model in question predicts a better performance than AMC, and a positive value indicates that AMC predicts a performance better than the model in question).

Our results show that a simple idle latency model can over-predict the performance by nearly 20% for an 8-core system. The performance difference is as little as 2% for a single-core system and

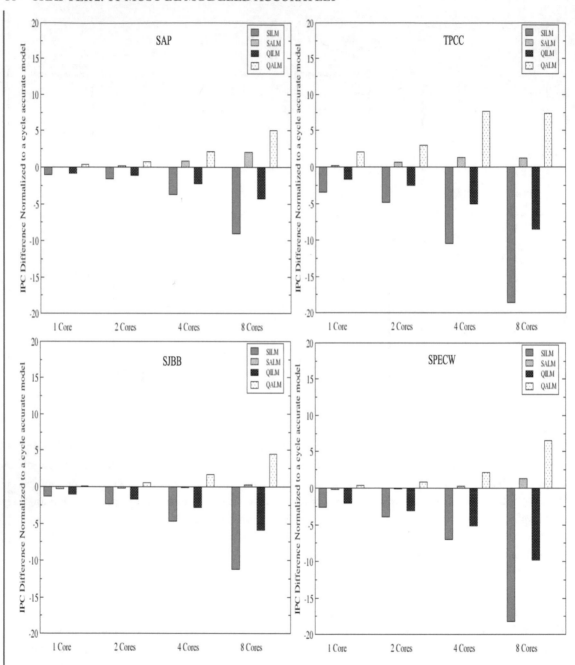

Figure 2.4: Performance comparison of various memory models. The x-axis shows the number of cores; the y-axis shows the performance difference for various memory models normalized to that of an accurate memory controller. We can observe that the performance difference increases with the number of cores.

increases steadily thereon. The memory bandwidth requirement increases with the number of cores on a chip, and as the number of cores increases, the system begins to operate in the exponential region of the bandwidth-latency curve (Figure 2.1). As Figure 2.1 shows, a fixed latency model does not capture the latency/bandwidth trade-off at all and hence fails to represent performance behavior accurately.

The queueing model, which does model a bandwidth constraint, fares better than the fixed latency models. The bandwidth sustained in this model is set to match that observed by the AMC (and so it changes on a per-benchmark basis). This scheme over-predicts the performance of the system by 10% for 8 cores. The performance difference is less than 2% for a single core system.

Note that the *SILM/QILM* models using the idle latency (the DRAM system's minimum request-turnaround time) consistently over-predict performance; that is, they under-predict the application execution time. The *SALM/QALM* models using the measured average latency of the benchmark in question tend to be less inaccurate in their predictions, and they tend to under-predict performance by over-predicting the application execution time.

Both the idle latency models (simple and queue-based) over-predict the performance of the system; the memory latency used in these simplistic models is the AMC's idle latency. We can observe from Figure 2.3 that the average latency in reality is between 1.3 to 1.4 times the idle latency of the system and depends on the bandwidth region in which the benchmark operates. In the exponential region of the bandwidth-latency curve, the expected request latency can easily be more than 3 times the idle latency of the system. Hence, these simplistic models over-predict the performance as they ignore this phenomenon. Further, the queuing model performs closer to AMC at higher bandwidth usage (larger number of cores) due to its modeling of bandwidth limitations.

The average memory models are more accurate than the idle latency models and always under-predict the performance of the system. This is due to the fact that most memory requests experience latency less than the average latency, i.e., the memory latency is unevenly distributed with respect to the average latency as shown in Figure 2.3 (the median is less than the mean, which is affected by a long tail). We noticed that almost 70% of requests experience a latency less than the mean value.

2.4   LET'S ADD PREFETCHING TO THE MIX

While it may appear that the preceding data suggests that what is needed is simply a more accurate simple model, for instance by using the median latency value rather than the mean or minimum latency value, there are two reasons to avoid following that conclusion and basing a simulation methodology on it:

- To find the actual median latency for a given benchmark on a given memory system, one must first perform the entire, accurate, detailed simulation to obtain that median value. The value changes on a per benchmark basis **and** on a per memory system basis.

- The results presented so far have been reasonably accurate (off by 20% at most) because the system we are modeling is simple: in-order CPUs that stall on cache misses. As the complexity of the system grows, so, too, does the inaccuracy of a simple model.

This section shows the effects of adding but a single dimension to the complexity of the system: prefetching.

Prefetching has been proposed as a powerful mechanism to reduce memory latency. Here we show the impact on the performance of various memory models when the system implements prefetching (Jouppi, 1990; Dahlgren and Stenstrom, 1995). The prefetching model under study uses a hardware stream prefetcher with a stream depth of 5. Each L3 cache miss (last level cache) initiates a memory request for the missing line (the requested line) and 5 subsequent predicted lines. The memory, cache and interconnect modules remained the same as the previous study. The cache misses and prefetch requests are given the same priority in the memory controller.

Figure 2.5 shows the performance difference of various memory models with respect to the AMC. The simplistic fixed-latency models behave similar to the earlier "no-prefetching" cases. The performance difference of simple idle latency model with AMC varies between 5% for single core systems to more than 60% for eight core systems. The higher bandwidth requirement of the prefetching scheme pushes the system to operate further into the exponential region of the bandwidth-latency curve, the zone where the simplistic models perform worse compared to AMC. The latency experienced by the requests is very high in this region, and having a low fixed latency without a bandwidth constraint will lead to erroneous conclusions. The queueing models also predict performance that is higher than for systems without prefetching; the performance difference of queuing models varies from 5% to 15% with respect to AMC.

Note that the models using average latency instead of idle latency (the *ALM results as opposed to the *ILM results) show an interesting behavior wherein the predictions get more accurate for the higher numbers of cores. This is a side-effect of an interesting real behavior. The average latency models first require a simulation to be run for that configuration, to obtain the correct average latency for that benchmark on that hardware configuration (i.e., the 8-core models use a different average latency than the 1-core and 2-core models, etc.). At eight cores, the system with prefetching is saturating the memory system to a degree that the difference between the mean and median latency values starts to decrease, which increases the accuracy of a model that uses the average memory latency.

2.5 SUMMARY

This section highlighted the drawbacks of using inaccurate/simplistic models. One of the main argument in favor of these simplistic models has been that they are sufficient to compute the performance difference between various systems, i.e., performance *trends*, though may not be useful for absolute values. Our studies show that these models can be wrong both in absolute performance numbers and relative performance comparison between different systems.

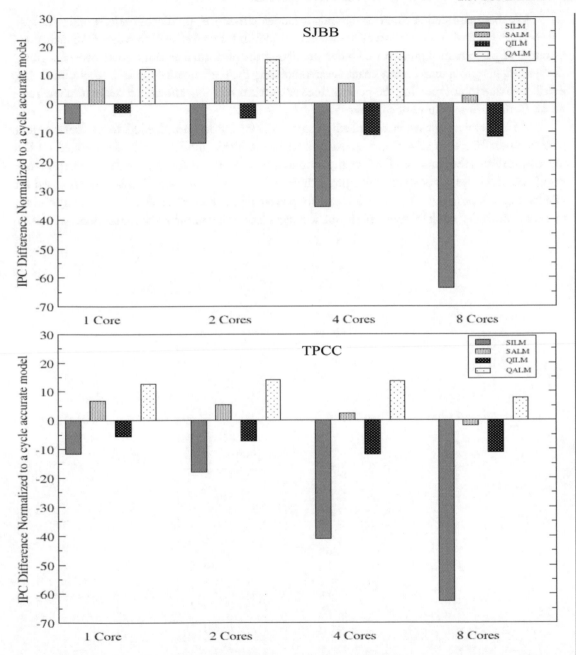

Figure 2.5: Performance comparison of various memory models with prefetching. This graphs shows the IPC difference normalized to an accurate memory controller for a multi-core CPU with a prefetching scheme [stream prefetcher with a prefetch depth of 5]. The performance difference between the simplistic models and AMC increases with the cores and is significantly greater than the no-prefetching scheme (cf. Figure 2.4).

We show case studies where the simplistic models either over-predict or under-predict the system performance with respect to a cycle-accurate model and therefore lead to wrongful conclusions in terms of performance projection. Under-predicting the performance can lead to over-designing the system, incurring unnecessary costs. Over-predicting the performance can lead to systems being ineffective due to not meeting the performance constraints of applications in a real world environment. Both cases are causes of concern.

As the system grows in complexity more accurate models are needed to evaluate system performance. The ease of use and speed offered by simplistic models are easily negated by the inaccurate results they produce. Further, the simplistic models are not able to capture the performance trends accurately as we observed with prefetching schemes. For instance, simple models based on idle latency will project performance increase for prefetching in server scenarios whereas an accurate memory controller model shows a performance degradation due to memory contention.

CHAPTER 3

... and It Will Change Soon

(life's tough, and then you go obsolete)

The future of memory systems promises to be interesting times ("interesting" in the sense of the pseudo-Chinese curse); it is not difficult to predict the occurrence of significant changes in the near future. It is a truism in computer-system design that all good ideas become bad ideas, and nearly all bad ideas become good ideas: the nature of the field is such that all ideas must be reevaluated in the light of any shift in technology, and computer-system technology details tend to change significantly every six months and radically every few years. Thus, every reevaluation rearranges the extant ideas in the field much as tossing a salad does to its ingredients.

Despite the fact that the memory-system industry is relatively averse to change (the DRAM industry in particular desires gradual, evolutionary changes), a number of current problematic trends point in similar directions for their solutions, and so it is quite possible that a disruptive solution will emerge victorious, and in the near term.

The bottom line: half of what I teach you today is likely to be worthless in five years. The trick is figuring out which half. Happy reading!

3.1 PROBLEMS AND TRENDS

The future of memory systems will likely be shaped by (our responses to) the following observations. The list is by no means exhaustive, but it is fairly representative of the important issues facing us today.

3.1.1 THE USE OF MULTIPLE CORES INCREASES WORKING-SET DE-MANDS

This is obvious in retrospect, but it certainly needs to be stated explicitly. As one adds cores to a system, one must also scale linearly both the bandwidth and capacity resources because a system with N simultaneously executing cores will have roughly N times the working set of a system with 1 core. Nothing comes for free. Several years back, the highly anticipated arrival of CMP was hailed by many as the *solution* to the memory-wall problem, rather than an exacerbation of the problem. Evidently, someone thought that having lots more work to do would enable CPUs to tolerate the limitations of the memory system much better. Needless to say, when CMPs did arrive in force, they straightened out that misconception relatively quickly. Nothing comes for free.

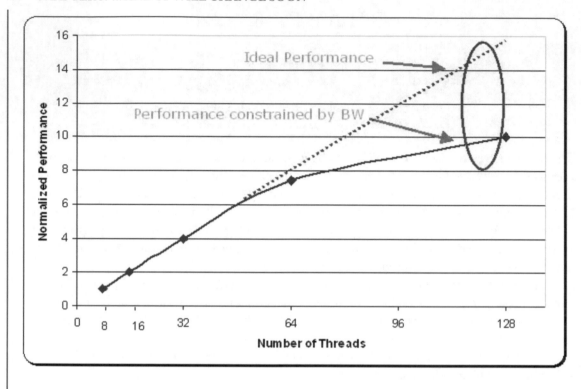

Figure 3.1: Performance of multicore vs. memory bandwidth. Figure taken from (Srinivasan, 2007).

3.1.2 MULTICORE BANDWIDTH REQUIREMENT IS ROUGHLY 1GB/S PER CORE

Related to the previous point: recent studies corroborate a general rule of thumb that the relationship between number of cores and bandwidth resources is somewhere in the neighborhood of 1GB/s. That is, as you add cores to a system, you will saturate your bus once the number of cores reaches somewhere around the maximum sustainable bus bandwidth divided by 1GB/s.

Figure 3.1 shows the results of a study done recently using IA32-compatible cores; it varies the workload, the number of cores (labeled "threads" in the graph) in the system, against a constant memory system having 52 GB/s *sustained* bandwidth (as opposed to peak theoretical bandwidth, which is simply the bus width times the bus speed). Performance scales linearly with the number of cores up to somewhere between 48 and 64 threads, at which point performance saturates. These results compare favorably with the fact that 32-core Niagara chips are shown to saturate at around 25 GB/s memory bandwidth.

An interesting point to note is that 2- and 4-core silicon has been available for several years in the commodity (e.g., desktop, low-end server) market, and 8-core has been well within reach for almost as long, but no significant commodity 8-core chips have appeared. High-end desktops and low-end servers have had 2- and 4-core CPUs for years; why have 8-core systems been so slow to arrive in this domain? One answer is that most machines in this domain use a single or dual DDRx memory system at datarates under 1Gbps per pin, which translates to a maximum sustainable bandwidth in the single-digit GB/s range. Meaning: one reason 8-core systems have not appeared is because, unless the memory system is upgraded (at significant cost), an 8-core system would perform no better than a 4-core system.

3.1.3   TLB REACH DOES NOT SCALE WELL (... OR AT ALL, REALLY)

Modern TLBs cannot handle the working sets thrown at them. The *translation lookaside buffer (TLB)* is a piece of hardware that allows the hardware to do quickly what the operating system does slowly: look an address up in the virtual memory system's page table to translate it from a virtual address to a physical address.

The concept of *TLB reach* (Talluri and Hill, 1994) is the amount of virtual space that a TLB can map: notionally, it is the number of TLB entries times the mapping power (corresponding page size) of each entry, and it represents the maximum amount of data/instructions application-level software can access at any given moment without causing an interrupt or trap. For instance, a 64-entry TLB with 4KB pages can map a total of 256KB; a working set larger than this will cause potentially numerous TLB misses. Think of it this way: you have a 16MB off-chip last-level cache, and your TLB can map only a fraction of it at a time; for instance with a 256KB reach, you need to take a TLB miss to access anything outside a small 1.5% of the cache's contents. The data is there in the cache, waiting to be accessed, but you can't touch it without causing a hiccup in processor performance.

The primary issue is that, while cache sizes have grown significantly in the last two decades, TLB sizes have not kept pace. Typical last-level caches in the late 1980s and early 1990s were tens to hundreds of kilobytes in size; typical last-level caches today are ones to tens of megabytes. The MIPS R2000 TLB, considered unthinkably large in the early 1990s, had 64 entries. A typical TLB today has anywhere from 256 entries (Pentium 4/M) to 1024 (G5 PowerPC). So caches have increased in their capacity by almost three orders of magnitude, while TLB reach has increased by just over a single order of magnitude at best. Of course, one can argue for superpages, which modern as well as past TLBs have supported, but a truly effective use of superpages is a tricky solution that should be as transparent as possible, requires a delicate interplay between application and operating system, and has not really been successfully demonstrated to date. Sun spent significant time working on the problem in the 1990s (Talluri et al., 1992; Khalidi et al., 1993; Talluri and Hill, 1994).

Historically, virtual memory has been seen as an acceptable overhead: it provides significant benefits (Jacob and Mudge, 1997, 1998a,c), and these come at a reasonable cost: Clark measured its overhead as roughly 10% in the 1980s (Clark et al., 1988; Clark and Emer, 1985), and we measured a similar overhead as late as 1998 (Jacob and Mudge, 1998b). It is important to note that previous studies (Clark's and our own) have been in the context of uniprocessors; the move to multicore seems to have changed things, which should not be surprising. Notably, a recent study shows the proportion of time spent handling TLB misses, which represents only one component of virtual memory overhead, is now 20% (Choi, 2008).

A significant obstacle is that, in contrast to the cache being off-chip, transparent to the operating system, and a design parameter under the control of the system designer (i.e., the same microprocessor can communicate with off-chip caches of different sizes), the TLB is on-chip, completely exposed to the operating system, and its size cannot be changed. Microprocessor designers would like to see a larger TLB, but their budgets are limited, and it is hard to argue for a significant increase in the TLB size: typical TLBs already burn significant power (the PowerPC gets away with its large size by being set associative, not fully associative as is typical).

3.1.4   YOU CANNOT PHYSICALLY CONNECT TO ALL THE DRAM YOU CAN AFFORD TO PURCHASE

Commodity DRAM today costs a few dollars per gigabyte in module/DIMM format. That means a low-range server system, costing under $5,000, could easily justify something on the order of 64 GB of memory (roughly $1,000 of DRAM). The problem: you cannot physically put 64 GB of DRAM into a single commodity system. As Figure 1.1 showed, while the datarates of DRAM channels have been increasing over time, the number of DIMMs per channel has *decreased* to enable those higher speeds. The result is that total channel capacity has been more or less flat for years.

This, by the way, is the capacity problem—the problem that FB-DIMM was invented to solve. There are many, many people out there who would love to put a huge handful of memory modules into their desktops or commodity blade servers; these people have the money to spend and the desire to spend it, but there is no low-cost solution that enables them to *install* the DRAM that they can afford to purchase.

At the higher end, the problem is somewhat solved: instead of a few dollars, one can pay a few *thousands* of dollars for the same capacity DIMM, but it is one that is engineered to support more than just two DIMMs per channel. Higher-end systems can put four DIMMs into a single DDRx channel, doubling the channel capacity. For this, people pay two orders of magnitude more money for the DRAM. The business model of several Silicon Valley startups and segments of some established companies as well exploits this window of opportunity, enabling lower-cost solutions for higher DRAM capacity, but, even so, we are not yet anywhere near the commodity-scale price-per-bit for server-scale capacity.

3.1.5   DRAM REFRESH IS BECOMING EXPENSIVE IN BOTH POWER AND TIME

The intuition behind this is straightforward: refresh power scales roughly with the number of bits in the system; refresh time scales roughly with the number of rows that need to be refreshed in a given time period. Refresh power is fundamentally used to charge up the leaky capacitors that are used to store the data. Even though these are actually *extremely* good capacitors, they do not hold charge forever. Moreover, the hotter they get, the more they leak, which is a significant issue as chips move to increasingly fast interfaces.

The time overhead of refresh scales roughly with the number and size of DRAM rows/pages in a system. DRAM rows have been in the several-thousand-bits size range for many years, whereas DRAM system capacity has grown enormously in the same timespan. For instance, a DRAM row was 4096 bits in 1990 (Prince, 1999); a typical row is 8192 bits today, almost two decades later (e.g., 1Gb x4 and x8 DDR3 parts). In 1990, a typical workstation had ones to tens of megabytes of DRAM; today's workstations have ones to tens of *gigabytes*.

Do the math, and you discover that a typical system in the 1990s had on the order of a thousand DRAM rows, each of which needed to be refreshed once every 64ms. That is on the order of 10 refreshes every millisecond. Today's systems have on the order of a million DRAM rows and use the same 64ms refresh period; thus you see about 10,000 refreshes every millisecond. To begin with, this represents a 1000-fold increase in refresh activity, which increases power dissipated in the DRAM system significantly. Further, consider that each refresh takes on the order of 100ns; this therefore represents about a millisecond of refresh activity every millisecond in a modern system. The only reason that systems are not completely incapacitated by refresh is the large number of independent banks in the system, enabling concurrent refresh. However, if the refresh activity is scheduled and controlled by the memory controller, all of those refresh commands must occupy the command bus: a single resource, and the refresh-command bandwidth is starting to impinge on performance.

3.1.6   FLASH IS EATING DISK'S LUNCH

This is perhaps obvious, but, again, it should be stated explicitly in a list of important trends: recent advances in flash memory (which seem largely related to volume pricing driven by applications such as handheld multimedia devices) have significantly reduced the price-per-bit gap between solid-state storage and hard drives. At disk's low-capacity end, the effect has been devastating; for example, Hitachi is considering shutting down its one-inch microdrive facility, once considered a shining jewel of the company.

It is also worth noting the other advantages flash holds over hard disk: its access time and power dissipation for writes are both closer to DRAM specs than disk, and flash's read power is actually lower than that of DRAM. The primary reason flash has not yet challenged DRAM for a main-memory technology is its limited write cycles: flash becomes unusable after it has been written too many times (on the order of 10^5 to 10^6 writes). While that limit is completely acceptable within the context of a file system, it is a relatively low number in the context of main memory.

3.1.7   FOR LARGE SYSTEMS, POWER DISSIPATION OF DRAM EXCEEDS THAT OF CPUS

As our own studies have shown (Jacob et al., 2007; Tuaycharoen, 2006), DRAM power dissipation for large installations is quite significant. In particular, FB-DIMM based systems can dissipate several hundred Watts per CPU socket: in the same number of pins as a dual-channel system, one can fit 6

FB-DIMM channels, each with 8 DIMMs, each dissipating 5–10W. Though much of that is in the interconnect chips (AMBs), which will be addressed with LR-DIMM (*Load Reducing DIMMs*, the next generation memory-system specification being discussed and designed within JEDEC), the fundamental limit remains: if you want more bits per CPU socket, be prepared to pay more power. It takes power to keep those bits alive (even in low-power sleep mode, the DRAM is still refreshing its data), so the only way to have tons of storage on the cheap is never to use it, which begs the question.

On second thought, begging the question is occasionally food for research: it is probably well worth exploring the design space to see if having more than enough DRAM storage is better than having just enough DRAM storage. Given that the alternative of going to disk is costly in both time and power, perhaps the excess capacity, even if kept in power-down mode most of the time, can provide more benefits than it costs.

3.1.8   ON-CHIP CACHE HIERARCHIES ARE COMPLEX

At the moment, there is no "best" cache design for on-chip caches (Jaleel and Jacob, 2006). They must satisfy several different requirements that pull a design in different directions. The on-chip cache must be low latency (despite PA-RISC's use throughout the 1990s of *very* large off-chip level-1 caches with long access times, most architects want a 1-cycle access to their L1 cache), it must have extremely high bandwidth (multiple 64-bit data readouts per sub-nanosecond clock cycle), and it must be extremely large (10s of MB isn't too much). To date, no cache design has been able to provide all these requirements in one package, and so many on-chip caches are on-chip cache *hierarchies*, with the L1 being small enough to provide single-cycle access, the L2 providing more capacity at a sustainable bandwidth, and so forth.

Tightly coupled to this is the constraint placed upon the cache system by the operating system, and an understanding of this constraint could perhaps explain why the PA-RISC example (using an L1 cache two to three orders of magnitude larger than those of other contemporary CPUs) seems such an outlier.

Simply put, the primary reason the L1 cache typically hovers around 32KB, and for most CPU designs has hovered there for the past two decades, is due to the operating system, specifically the virtual memory subsystem. The most common operating system, Windows, assumes a physically indexed cache, a design decision that solves the virtual address aliasing problem [Jacob et al. (2007), "Section 4.2.1, Virtual Cache Management"]. This is a non-trivial problem: not in that it is difficult to solve but in that it *must* be addressed if a modern computer is to work correctly, and the most commonly used solutions come at great costs.

For instance, the solution used by Windows—the assumption of a physical cache in which aliases cannot occur—implies that the fundamental cache unit, the number of sets times the cache block size, cannot be larger than a page size. The fundamental cache size is illustrated in Figure 3.2; it depends upon the virtual memory system because the number of virtual pages that fit into this space indicates the number of aliases that a physical datum can have in a virtually indexed cache.

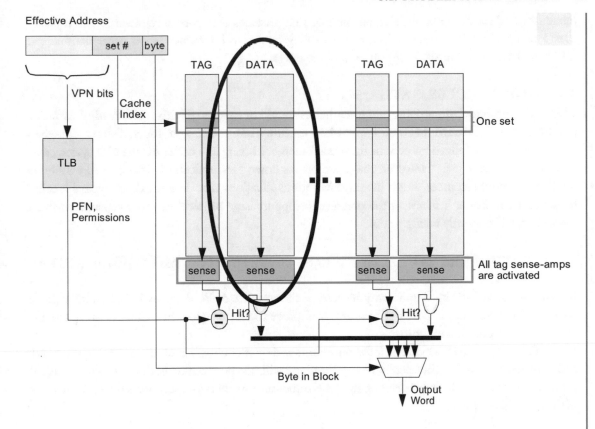

Figure 3.2: The fundamental cache size. This is the number of sets (usually 2^n, where n is the index size) times the cache block size. The figure shows an n-way set associative cache with the fundamental unit of storage circled. This size is important because this is the space into which an address maps, and if this size is larger than a page, a particular physical datum could map to multiple sets in a virtually indexed cache.

Thus, if an operating system expects a physically indexed cache (implying that it does no extra effort to address the aliasing problem), and the cache uses a virtual index (as shown in the figure), then this fundamental cache unit must be no larger than a virtual page. Given such a requirement, the only way to create a large cache is by increasing associativity, and so it should not be surprising at all that so many CPUs designed for the PC market have had configurations such as 4KB, 8KB 2-way associative, 16KB 4-way set associative, 32KB 8-way associative, etc.

The PA-RISC was not intended to run commodity operating systems but rather Hewlett-Packard's flavor of Unix, and HP designers were well aware of the aliasing problem (Huck and Hays, 1993). Thus, the PA-RISC was not designed to solve the aliasing problem in the same way as PC-oriented CPUs. In the mid-1990s, it was intended to run large commercial applications such as databases, and the designers insisted on extremely large caches at the first level, focusing on capacity over latency. PA-RISC systems routinely had megabyte-size L1 caches when other commercial CPUs had L1 caches in the single-digit kilobyte range.

3.1.9   DISK ACCESS IS STILL SLOW

This is yet another obvious point that nonetheless should be made explicitly: system administrators purchase enough DRAM that their applications execute without having to touch disk (except when necessary). Most applications benefit from main memory acting as a cache for the disk system, and so this tactic works well. However, the tactic breaks down for applications that exceed a system's physical memory resources, and it breaks down for applications that do not exhibit typical locality behavior. In addition, it increases the pressure to support large physical memory capacities, which the industry is currently failing to do.

3.1.10   THERE ARE TOO MANY WIRES ON A TYPICAL MOTHERBOARD AS IT IS

Increasing bandwidth by significantly increasing the number of PCB traces is not really a viable option. Moreover, any proposed solution must improve the status quo while using existing specifications, architectures, and protocols.

This argument transcends just the pin counts and trace counts; it extends to existing interfaces and standards as well; a completely new technology will not be received with open arms unless it promises at least a 10x improvement in some dimension. Otherwise, any newcomer must work within existing bounds.

3.1.11   NUMEROUS NEW TECHNOLOGIES ARE IN DEVELOPMENT

There are numerous new technologies in active development, including ZRAM, FeRAM, MRAM, PRAM (phase change), and others. Of course, each has only a small chance of making it to the mass market and having a significant impact, but the sheer number of upcoming novel solid-state storage technologies being explored makes it quite likely that at least one will do so.

3.2   SOME OBVIOUS CONCLUSIONS

Given the preceding list of issues and observed trends, what could one conclude? There are a handful of obvious implications (with detail to follow):

- A new DRAM organization is needed, one that supports the bandwidth needs of multicore CPUs, takes the pressure off the TLB, and supports 100s of GB of commodity DRAM modules.

- Flash needs to be integrated into the memory hierarchy as a first-class citizen—i.e., more so than a mere removable thumb drive.

- It is probably worth re-thinking the implementation details of superpages, because they certainly do address some problems of overhead.

3.2.1 A NEW DRAM-SYSTEM ORGANIZATION IS NEEDED

Two of the most pressing problems can really only be solved effectively and inexpensively by a reorganization of the DRAM system: those problems being support for multicore (which suggests significant concurrency, equal to or greater than FB-DIMM) and a significant increase in capacity at the commodity level. Coupled with this is the requirement that any new organization must use existing standards; e.g., it must resemble either DDRx SDRAM or FB-DIMM in the technologies it chooses, else it has virtually no chance of (immediate) adoption.

The closest solution at hand is FB-DIMM, which supports significantly more concurrency and storage than a DDRx system, given an equal number of controller pins and motherboard traces. FB-DIMM also requires significantly more power than the commodity market is willing to subsidize, so without at least a minor re-design, even it is not the ultimate solution.

3.2.2 FLASH NEEDS TO BE INTEGRATED

Since it offers things no existing solid-state technology does and now does so at price points that are increasingly tempting, it is well worth investigating how flash memory can be used as a first-class citizen within the memory hierarchy, rather than as a simple removable disk drive. For one thing, integrating flash memory as a permanent resource can potentially eliminate the overhead of accessing its storage through the operating system, which is significant and necessary if flash appears as a disk drive. The one weak point of flash, its limited write capability, it significant but must be characterized within a real system model (e.g., Would a power user be willing to change out flash DIMMs once per year if the pay-off was a doubling in system performance? Probably. Would an average user? Time will tell).

Computers have always exploited special-purpose hardware for special-needs applications, when the payoff warranted the expense. Flash offers low-power reads (for read-mostly data), non-volatile storage with low power and low access-time specs (relative to disk), solid-state construction (i.e., decent shock resistance), and a compelling form factor. This is definitely special-purpose hardware; the only questions are what applications can best exploit its capabilities? and do the benefits justify the extra costs?

3.2.3 POSSIBLY REVISIT SUPERPAGES

One of the costs exposed by increased TLB overhead (i.e., increasing TLB misses due to relatively flat TLB sizes against working-set sizes that increase over time) and increasing disk-transfer time is that of data aggregation: namely the limitation of the 4KB page size. A typical TLB maps a pitiful amount of space, and the high cost of going to the disk to transfer a page is amortized over what is a very small amount of data that is getting smaller relative to nearly everything else (photos are getting larger, videos are getting larger, all documents from email to SMS are incorporating multimedia, etc.—4KB is nothing compared to that).

Superpages have been investigated in the past to solve this problem, the idea being that some pages should be much larger than others but still correspond to a single TLB entry and a contiguous range of addresses in both virtual and physical memory spaces. The mechanism has not found widespread acceptance at the operating systems level, largely due to a hard-wired support for "regular" pages in most operating systems that can only be circumvented through "substantial, invasive" modifications to the operating system (Talluri et al., 1992; Khalidi et al., 1993; Talluri and Hill, 1994).

This solution should not be discounted but rather investigated more intensely, as a successful implementation of superpages could address several current problems, including TLB reach and disk-access overhead.

3.3   SOME SUGGESTIONS

This final section gives concrete design suggestions that address many of the issues raised earlier. For instance:

- A spin on FB-DIMM that could increase capacity and concurrency and simultaneously reduce power dissipation to that of a DDRx system

- Some uses for flash that could improve performance and/or power dissipation at the system level

- A revisitation of superpages that might overcome previous barriers

- A new cache design that enables very large L1 caches

These are ideas that have been in development in our research group over the past decade.

3.3.1   FULLY BUFFERED DIMM, TAKE 2 (AKA "BOMB")

In the near term, the desired solution for the DRAM system is one that allows existing commodity DDRx DIMMs to be used, one that supports 100 DIMMs per CPU socket at a bare minimum, and one that does not require active heartbeats to keep its channels alive—i.e., it should resemble DDRx SDRAM in that power should only be dissipated for those DIMMs actively executing commands and reading/writing data. Currently, nothing like this exists in the commodity domain, but industry is pursuing something similar (but not quite as aggressive) in the LR-DIMM (load reducing DIMM) proposal. We call our design Buffer-on-MotherBoard.

An obvious organization that would provide these specifications is shown in Figure 3.3. Rather than placing the slave memory controllers on the DIMM, as is done in FB-DIMM, one can instead attach a master memory controller a large number of JEDEC-style DDRx subsystems, with the speed grades of the subsystems chosen to support a minimum of four DIMMs per channel. The

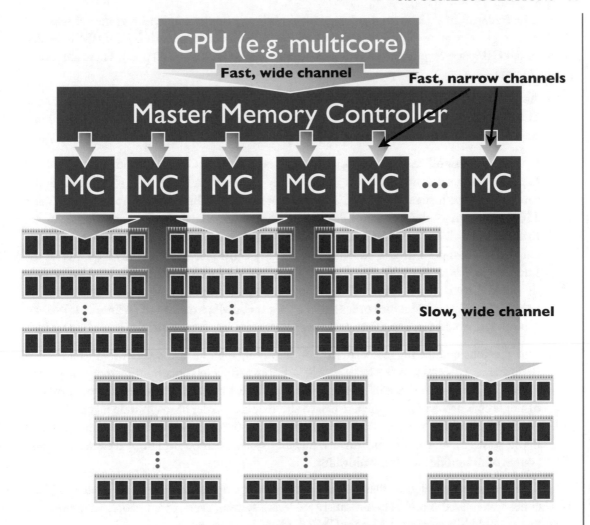

Figure 3.3: A DRAM-system organization to solve the capacity and power problems.

connections between the master MC and the slave MCs must be extremely low pin count (e.g., no more than 10 total pins per channel), to enable something like 64 individual sub-channels on the master memory controller, given a pin limit in the ~500 range. If each channel were 8 pins, and each sub-MC could handle four DIMMs, such an organization would support 256 DIMMs per CPU socket at a cost just over 500 pins. At that point, the problem would be figuring out how to cram 256 DIMMs into a standard chassis, arguably a good problem to have given the order-of-magnitude increase in capacity it enables.

The trade-off in such an arrangement would obviously be the bandwidth at the sub-channel level, which, given a mere eight pins would only be in the 1GB/s range (cf. FB-DIMM, which achieves 6.4 GB/s in ~65 pins, and XDR, which achieves 9.6 GB/s in ~100 pins). There are several points to note:

- As mentioned earlier, 1GB/s is a reasonable per-core bandwidth (though, to be fair, cores need 1GB/s *sustainable* bandwidth, not theoretical max bandwidth, and no channel sustains more than 75–80% of its theoretical max, so perhaps 1.3GB/s is a better target).

- If there is no significant bandwidth mismatch between the CPU side and DRAM side of the slave memory controllers, this will probably enable even more than four DIMMs per subchannel. For instance, a 1.2GB/s DDRx system is a 64-bit datapath running at a mere 150Mbps per pin—the datarate of PC150 SDRAM. An interesting twist to the design, rather than having a slow 64-bit datapath, would be slightly non-standard DIMMs with multiple chip-select lines per DIMM, enabling the memory controller to select only a quarter of the DRAMs in a rank to respond to a query. This would reduce power dissipation significantly (it would drive only a fraction of the bitlines and sense amps of a normal request), and, for example, the data could return over 16 of the lines at 600Mbps instead of 64 lines at 150Mbps. However, it would most certainly make supporting chipkill more difficult (if not impossible).

- Our previous studies suggest that, despite intuition, such an arrangement will have good performance (Cuppu and Jacob, 1999, 2001). Given sufficient concurrent memory operations in the CPU, a system made of multiple, low-bandwidth channels can equal the performance of single channels with an aggregate bandwidth that is an order of magnitude higher. In other words, bandwidth is not the whole story; how you use that bandwidth is more important, and if you waste it, your high-performance memory system can be beaten by a better designed system with significantly less bandwidth.

Our previous study found that a single out-of-order core can keep 2–4 channels busy, so a 64-channel system would probably not satisfy a 64-core system, even if those cores were in-order. However, it would likely satisfy a 32-core CMP made of in-order cores.

The power dissipation of this system would be significant during bursts of activity, but during periods of relatively little traffic, it would dissipate far less power than a FB-DIMM system, since there is no need to keep a daisy chain of AMBs passing timing heartbeats. Individual DIMMs would only dissipate power when actively servicing requests.

It is worth noting that this organization is not at all novel; such hierarchical arrangements of memory controllers have been implemented at the high-end server and supercomputer levels for years [e.g., (Hotchkiss et al., 1996; Bryg et al., 1996)]. Moreover, nearly all of today's high-profile technologies were originally developed for supercomputers (for example caches, pipelining, out-of-order execution, etc.) and ultimately found their way into the commodity market as the needs arose and as the costs for adoption decreased. High-performance memory-system design is just another such technology.

3.3.2   SOME USES FOR FLASH

Flash is a household word; there is no question of its success as a semiconductor memory technology. Even more, it has the potential to become a disruptive technology within the computer system, because it enables non-linear jumps in computing performance and capabilities. For example, our research (Jacob et al., 2007) shows that one can free up the operating system and its use of DRAM when heavily loaded, to the tune of a 100% increase in effective capacity, by using flash to buffer disk writes. In other words, through the effective use of flash, one can put roughly twice the number of executing threads onto a system without having to double the DRAM storage. A factor of two is significant, especially at a time when researchers scramble for 5–10% gains using existing technologies.

The primary enabling step is to divorce flash from its disk-drive interface; the disk-drive interface requires a file system and thus an OS system call for access. Going through the operating system for a memory access has tremendous overhead (imagine removing the TLB and translating all virtual memory addresses through the operating system and its page table); flash would be best accessed as we currently access "normal" solid-state memory—i.e., through the virtual memory system and thus the TLB. This arrangement would reduce the cost per bit of flash memory to the same level as that of a large SSD, as the cost of the controller would be spread out across potentially multiple DIMMs of flash; a single flash controller would be responsible for buffering writes, identifying free blocks, and garbage collecting for multiple flash DIMMs in a channel. Leaders in the semiconductor memory industry are already developing this idea.

There are obvious uses for such a mechanism: as mentioned, it makes a good buffer for the disk system (though not as transparent as placing the buffer on the disk side), and for embedded systems and smaller general-purpose systems it can obviously replace disk entirely. Other studies have shown the performance benefit of battery-backing the DRAM system (Chen et al., 1996), achieving better system-level behavior when you know you can count on the integrity of your main memory, which is directly applicable.

In addition, the operating system can exploit differentials in energy/performance and segregate items that have different access characteristics, as is currently done in embedded systems today but with memory technologies such as scratch-pad SRAM vs. DRAM vs. ROM. For instance, consider that flash has a lower read energy than DRAM: it takes less energy to read a cache block of data from flash than it does to read it from DRAM. One could exploit this differential by placing read-mostly data into the flash channel and write data into the DRAM channel—e.g., segregate normal data from instructions and constants, mapping normal data into the DRAM subsystem and mapping instructions and constants into the flash subsystem. All of this would be hidden from the application by the indirection of the virtual memory system. More sophisticated heuristics are clearly possible, and much of this terrain has already been explored by the embedded-systems community.

3.3.3 SUPERPAGES (TAKE 2) AND SUPERTLBS

It is time to revamp virtual memory from the hardware side, as it is clear that the problem will not be addressed from the operating-system side (as Talluri and Hill (1994) imply, one cannot expect OS rewrites to solve one's problems). In particular, the concept of superpages needs to be revisited, since most hardware architectures support superpages, despite the fact that most operating systems do not (at least, not beyond the obvious uses such as mapping special kernel structures). The good news is that this can be done readily in a combined hardware/virtual-machine approach, without support from (or the approval of) the operating system.

The first point to exploit is that we are now (and have been for some time) in the age of cheap memory. DRAM costs very little more than disk at the bit level when you consider its performance advantage. One can afford to be wasteful with it, provided the system has enough capacity—i.e., the limitation is not the wallet but the system engineering (see discussions above). Given enough capacity, one can afford to allocate superpages and fill them sparsely—a 1MB superpage could be subblocked into smaller chunks, for instance into eight 128KB pieces, and each superpage translation entry would have an associated (in this case 8-bit) subblock-validity vector.

This second point to exploit is that TLBs, like caches, can be multi-level, and a second-level TLB can afford to have both a longer access time and a longer refill time than the first-level TLB. Thus, it pays to make an L2 TLB really, really large—in particular, it should be large enough to map *all* of the caches (e.g., 16MB TLB reach, requiring some 4000 entries), plus some additional room for growth, traffic spikes, mapping conflicts, etc. Call it 8000 entries, multiplied by whatever factor is needed to get the hit rate of a fully associative organization.

The third point to exploit is that VMMs (Smith and Nair, 2005) have become pervasive today, and their performance overhead is similar to that of virtual memory; i.e., perfectly acceptable, given the benefits they provide. A VMM can (and frequently does) perform re-mapping of virtual addresses out from underneath the operating system (Accetta et al., 1986). This creates an additional level of address mapping: the operating system's concept of physical memory becomes virtual. If done right, this can enable the performance benefit of superpages without actually requiring any operating-system support: the VMM is in a position to monitor activity and aggregate hot pages (at the subblock granularity) into superpages which can then be mapped by a single L1 TLB entry. Talluri and Hill outline one approach (Talluri and Hill, 1994) that would translate well to a VMM implementation. This can also be combined with the "hot-spot vector" concept of remapping pages for cache performance (Jacob et al., 2007).

3.3.4 THE HASH-ASSOCIATIVE CACHE

As mentioned earlier, cache design is still an open problem: a processor cache must address the issues of capacity, bandwidth, and latency, all within the constraints of the virtual memory system and the potential for virtual address aliasing. The vast majority of L1 processor caches are set associative, with the fundamental cache size being that of the virtual page, a design chosen to support the legacy operating-system design decision of assuming physically indexed caches. To increase capacity of the

L1 cache, given this design choice (a virtually indexed cache masquerading as a physically indexed cache), one must keep the fundamental cache size fixed and increase the degree of associativity.

Among other things, the choice begs the question of power dissipation. Power dissipation is a primary design constraint that affects all manner of computer systems, from hand-held to high-performance, and within a given design, processor caches are typically one of the most significant contributors to both static and dynamic power dissipation. Traditional set-associative caches are unattractive for use as level-1 caches because they dissipate more power than a direct-mapped cache of similar performance. Nonetheless, traditional set-associative caches are commonly used for level-1 data caches in many modern designs (e.g., the latest x86-compatible, PowerPC, SPARC, and Itanium architectures), largely because of the aliasing problem: traditional set-associative cache presents a smaller set-index to the system than a direct-mapped cache of the same capacity, thereby reducing (to the point of eliminating) the system's exposure to the virtual-address aliasing problem.

This is clearly not the optimal design choice; note well that most L1 caches have stopped at 8-way set associativity, despite the need for much, much larger caches. What a designer needs is the ability to make a much larger L1 cache, virtually indexed and pretending to be physically indexed, without the power-dissipation increase commensurate with the traditional set-associative cache.

Enter the *hash-associative cache*, an architecture-level design that combines the concepts of banking and hashed indices to produce a highly set-associative cache with the power dissipation and access time of a highly-banked direct-mapped cache and yet the performance (in miss rate, CPI, and total execution time) of an NMRU-managed set-associative cache. Providing a high degree of associativity allows one to build a large level-1 cache that nonetheless eliminates the virtual-address aliasing problem. The fundamental idea is to eliminate the n-way tag comparison of an n-way set-associative cache by using a hash function to preselect one cache way—or two or more ways, if a multi-probe design is desired. The fundamental cache unit is still equal to a virtual page size, so the design eliminates virtual address aliasing, and the deterministic way-lookup enables the size to be scaled far beyond a traditional n-way set-associative cache; the size limit becomes effectively the same as a banked direct-mapped cache.

The architecture is shown in Figure (3.4) with comparable organizations of an n-way set-associative cache and an n-way banked direct-mapped cache. The traditional n-way set-associative cache drives the sense amps for n tag arrays and makes n tag comparisons. The direct-mapped and hash-associative caches each drive one set of sense amps and make one tag comparison.

A detail to consider is the replacement of a set's most-recently-used block. If a single probe is used, the cache will dissipate the smallest power, but if the identified way contains the most recently used block in the set, then that block would be replaced on a cache miss; this is not desirable. A multi-way probe would solve this problem (e.g., using the complement of the hash, or even multiple hashes), at the expense of increased power dissipation; a two-probe implementation would dissipate roughly the same power as a 2-way set-associative design of the same size.

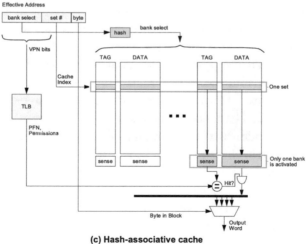

Figure 3.4: The hash-associative cache. The organization and operation of a hash-associative cache is compared to that of a traditional *n*-way set-associative cache (a) and a traditional *n*-way banked direct-mapped cache (b). In particular, note that the traditional *n*-way set-associative cache drives the sense amps for *n* tag arrays and makes n tag comparisons. The direct-mapped and hash-associative caches each drive one set of sense amps and make one tag comparison.

3.4   VIRTUAL MEMORY IN THE AGE OF CHEAP MEMORY

Virtual memory was originally invented to automate paging. However, we have reached a point at which paging is not the norm but the exception—paging is an aberration that determines a system's capacity for performance, i.e., you execute as much stuff as you can until you start to page, and that is your ceiling.

So why not just admit it?

Virtual memory has outlived its original purpose; one could even argue it has outlived much of its usefulness. Along the way, it has become operator-overloaded with numerous additional functions; the virtual memory system now wears numerous hats, not all of them compatible with each other. Nonetheless, the mechanism provides for a ludicrously long list of features that users have come to expect of modern systems (Jacob and Mudge, 1997):

- Address space protection and multitasking

- Shared memory, in particular secure shared memory

- Large address spaces and the automated movement of data from DRAM to disk

- Fine-grained protection (e.g., read/write/execute at the page level)

- The fragmentation and partial loading of an application's executable image (i.e., an app need not be contiguous in physical memory or entirely memory-resident to execute)

- Support for sparse address spaces (e.g., dynamically loaded shared libraries, multiple threads, etc.)

- Direct memory access

- Memory-mapped files and I/O

Why not eliminate some of these features, or provide them via some other mechanism besides demand-paged 4K-page-granularity virtual memory? For instance, exploit large, cheap physical memory sizes and *try* to have large contiguous ranges of instructions and data; don't worry about large unused holes, as they might fill up later as applications' working sets change. Use the virtual memory system to do partial loading of applications, but (a) dramatically increase the granularity of data movement to/from the disk, and (b) don't worry so much about using the memory space efficiently. For decades, the physical memory system has been organized as an associative cache for the disk system, and this design choice was made when bits were precious. They are no longer. Revisit the design in light of this fact.

A related issue is that of direct memory access and multiprocessing, namely that there are multiple entities in the memory system that all want to access the same object but use potentially different names for it. The DMA engine uses physical addresses; other processors use (potentially) different virtual addresses for the same physical location. This creates naming problems that are often solved with brute-force methods (e.g., cache flushing, TLB flushing and shootdown, physically indexed caches, etc.). Why not move the translation point out further (e.g., put the TLB out near the memory controller) and let all of the entities involved occupy the same namespace?

Protection and multitasking can be accomplished by PowerPC/PA-RISC-style segments, as can shared memory, memory-mapped files & I/O, and even "fine" grained protection (if one is willing to live with a larger unit of granularity). Large address spaces are no longer an issue; application datasets are only larger than the physical memory for select applications. The only features left that are worthwhile and uniquely suited to virtual memory and its page-granularity mapping/data-movement ability are ... um, nothing?

Postlude: You Can't Fake It

(it isn't *really* the age of cheap memory)

Even though DRAM costs a dollar per gigabit, the memory system is really in no way cheap, because it dictates system-level performance, and in large installations it can dictate the system's power dissipation and thermal envelope as well.

The cost of the memory system is not in its bits—that is misleading. However, the community has largely been lulled into thinking that way, that memory is cheap due to its dollar cost per bit, and therefore the community has paid it little heed over the past few decades. This is one of the reasons that the memory system has become and remained the bottleneck that it now is and has been for some time.

The cost of the memory system is not in the bits but in the design. Meaning: if you do not design it well, you pay the price of significantly reduced performance and/or significantly increased power dissipation. As has been shown, the memory system design space is highly non-linear, and the delta in power and execution time between good configurations and bad configurations can be an order of magnitude (Cuppu and Jacob, 2001; Tuaycharoen, 2006; Jacob et al., 2007). It is not an issue of 10% ... it is an issue of 10x.

There are scheduling issues at multiple levels within the memory system; there are numerous choices for heuristics; the design space is enormous, sparsely explored, frequently misrepresented, and consequently poorly understood. The only way to explore the design space and/or evaluate the heuristics is to rigorously characterize the entire system via detailed, highly accurate simulation using real benchmarks. In the context of today's relatively complex systems, statistical models of the system fail; statistical models of the heuristics fail; statistical models of workloads fail. Simplistic models fare little better.

The only way to determine where to go next is to do system-level simulation. Because many of the issues directly involve the operating system, it must be included in the simulation framework, which requires high accuracy, deep simulation models, and highly detailed frameworks. Nothing comes for free. This is a complex beast, at least as complex as a high-performance CPU; it requires the same attention to detail.

ACRONYMS

ACT — activate (DRAM command)

AMB — advanced memory buffer (FB-DIMM's interface chip)

AMMP — a benchmark application

ART — a benchmark application

CAS — column address strobe (DRAM signal)

CPI — cycles per instruction (a metric for performance, inverse of IPC)

CPU — central processing unit (i.e., microprocessor)

DDR — double data rate (a type of SDRAM)

DDR1, DDR2, DDR3, etc. — different generations of DDR technology (DDRx in general)

DIMM — dual in-line memory module (typical unit of DRAM storage for systems)

DMA — direct memory access (a high-performance mechanism)

DRAM — dynamic random-access memory (a solid-state memory technology)

FB-DIMM — fully buffered DIMM (a type of DIMM as well as memory-system design)

FBD — shorthand for FB-DIMM

FeRAM — ferro-electric random-access memory (a non-volatile solid-state memory technology)

FIFO — first in, first out (queue)

IPC — instructions per cycle (a metric for performance, inverse of CPI)

JEDEC — joint electron device engineering council (standards organization)

MC — memory controller (controls DRAM operation)

MRAM — magnetoresistive random-access memory (a non-volatile solid-state memory technology)

NMRU — not most recently used (a cache-management heuristic)

PA-RISC — performance architecture RISC (a Hewlett-Packard CPU design)

PCB — printed circuit board (on which multiple chips are integrated)

PRAM — phase-change random-access memory (a non-volatile solid-state memory technology)

PRE — precharge (DRAM command)

PSRAM — pseudo-static random-access memory (a non-volatile solid-state memory technology)

RAS — row-address strobe (DRAM signal)

RAID — redundant array of inexpensive, later changed to *independent*, disks (a storage technology)

RAM — random-access memory (a classification of storage technology)

RD — CAS read (DRAM command)

RLDRAM — reduced latency DRAM (a solid-state memory technology)

ROM — read-only memory (a classification of solid-state memory technology)

SAP — a benchmark application

SDRAM — synchronous DRAM (a classification of DRAM)

SJBB — a benchmark application

SO-DIMM — small outline DIMM (a type of DIMM for laptops)

SPARC — scalable processor architecture (a Sun CPU design)

SPEC — standard performance evaluation corporation (benchmark suite & company)
SPECW — a benchmark application
SRAM — static random-access memory (a solid-state memory technology)
SSD — solid-state disk (a solid-state memory technology)
TLB — translation lookaside buffer (hardware to support operating-system functions)
TPCC — a benchmark application
VMM — virtual machine manager (sub-operating-system HW/SW layer)
WR — CAS write (DRAM command)
ZRAM — zero-capacitor random-access memory (a non-volatile solid-state memory technology)

Bibliography

M. Accetta, R. Baron, W. Bolosky, D. Golub, R. Rashid, A. Tevanian, et al. (1986). "Mach: A new kernel foundation for UNIX development." In *USENIX Technical Conference Proceedings*, pp. 93–112.

W.R. Bryg, K.K. Chan, and N.S. Fiduccia (1996). "A high-performance, low-cost multi-processor bus for workstations and midrange servers." The *Hewlett-Packard Journal*, 47(1). DOI: 10.1145/237090.237154

P.M. Chen, W.T. Ng, S. Chandra, C. Aycock, G. Rajamani, and D. Lowell (1996). "The RIO file cache: Surviving operating system crashes." In *Proc 7th Int'l Conference on Architectural Support for Programming Languages and Operating Systems*, pp. 74–83.

I. Choi (2008, July). *Personal communication.*

D.W. Clark and J.S. Emer (1985). "Performance of the VAX-11/780 translation buffer: Simulation and measurement." *ACM Transactions on Computer Systems*, 3(1), 31–62. DOI: 10.1145/214451.214455

D.W. Clark, P.J. Bannon, and J.B. Keller (1988). "Measuring VAX 8800 performance with a histogram hardware monitor." In *Proc. 15th Annual International Symposium on Computer Architecture (ISCA'88)*. DOI: 10.1145/633625.52421

V. Cuppu and B. Jacob (1999). "Organizational design trade-offs at the DRAM, memory bus, and memory-controller level: Initial results." Technical Report UMD-SCA-1999-2, University of Maryland Systems & Computer Architecture Group.

V. Cuppu and B. Jacob (2001). "Concurrency, latency, or system overhead: Which has the largest impact on uniprocessor DRAM-system performance?" In *Proc. 28th Annual International Symposium on Computer Architecture (ISCA'01)*, pp. 62–71. Goteborg, Sweden. DOI: 10.1109/ISCA.2001.937433

V. Cuppu, B. Jacob, B. Davis, and T. Mudge (1999). "A performance comparison of contemporary DRAM architectures." In *Proc. 26th Annual International Symposium on Computer Architecture (ISCA'99)*, pp. 222–233. Atlanta GA. DOI: 10.1145/300979.300998

V. Cuppu, B. Jacob, B. Davis, and T. Mudge (2001). "High performance drams in workstation environments." *IEEE Transactions on Computers*, 50(11), 1133–1153. DOI: 10.1109/12.966491

F. Dahlgren and P. Stenstrom (1995). "Effectiveness of hardware-based stride and sequential prefetching in shared-memory multiprocessors." In *First IEEE Symposium on High-Performance Computer Architecture, 1995. Proceedings*, pp. 68–77.

P.G. Emma, A. Hartstein, T.R. Puzak, and V. Srinivasan (2005). "Exploring the limits of prefetching." *IBM Journal of Research and Development*, 49(1), 127–144.

B. Ganesh (2007). *Understanding and Optimizing High-Speed Serial Memory-System Protocols*. Ph.D. Thesis, University of Maryland, Dept. of Electrical & Computer Engineering.

T.R. Hotchkiss, N.D. Marschke, and R.M. McColsky (1996). "A new memory system design for commercial and technical computing products." *The Hewlett-Packard Journal*, 47(1).

J. Huck and J. Hays (1993). "Architectural support for translation table management in large address-space machines." In *Proc. 20th Annual International Symposium on Computer Architecture (ISCA'93)*, pp. 39–50. DOI: 10.1145/173682.165128

I. Hur and C. Lin (2008). "A comprehensive approach to DRAM power management." In *Proc. International Symposium on High-Performance Computer Architecture*, pp. 305–316. Austin TX. DOI: 10.1109/HPCA.2008.4658648

B. Jacob, S. Ng, and D. Wang (2007). *Memory Systems: Cache, DRAM, Disk*. Morgan Kaufmann Publishers Inc. San Francisco CA, USA.

B. Jacob and T. Mudge (1997). "Software managed address translation." In *Proc. Third International Symposium on High Performance Computer Architecture (HPCA'97)*, pp. 156–167. San Antonio TX. DOI: 10.1109/HPCA.1997.569652

B. Jacob and T. Mudge (1998a). "Virtual memory: Issues of implementation." *IEEE Computer*, 31(6), 33–43. DOI: 10.1109/2.683005

B. Jacob and T. Mudge (1998b). "A look at several memory-management units, TLB-refill mechanisms, and page table organizations." In *Proc. Eighth Int'l Conf. on Architectural Support for Programming Languages and Operating Systems (ASPLOS'98)*, pp. 295–306. San Jose CA. DOI: 10.1145/291069.291065

B. Jacob and T. Mudge (1998c). "Virtual memory in contemporary microprocessors." *IEEE Micro*, 18(4), 60–75. DOI: 10.1109/40.710872

A. Jaleel and B. Jacob (2006). "Using virtual load/store queues (VLSQ) to reduce the negative effects of reordered memory instructions." In *Proc. of the 11th Intl. Symp. on High Performance Computer Architecture*, pp. 191–200. San Francisco CA. DOI: 10.1109/HPCA.2005.42

JEDEC (2007). *JEDEC Standard JESD 82-20*: FBDIMM Advanced Memory Buffer (AMB).

N.P. Jouppi (1990). "Improving direct-mapped cache performance by the addition of a small fully-associative cache and prefetch buffers." In *Proc. 17th Annual International Symposium on Computer Architecture (ISCA'90)*, pp. 364–373. DOI: 10.1109/ISCA.1990.134547

R.E. Kessler and M.D. Hill (1992). "Page placement algorithms for large real-indexed caches." *ACM Transactions on Computer Systems (TOCS)*, 10(4), 338–359. DOI: 10.1145/138873.138876

Y.A. Khalidi, M. Talluri, M.N. Nelson, and D. Williams (1993). "Virtual memory support for multiple page sizes." In *Proc. Fourth Workshop on Workstation Operating Systems*, pp. 104–109. DOI: 10.1109/WWOS.1993.348164

L. Kleinrock (1975). *Queueing Systems, volume I: Theory*. John Wiley & Sons New York.

M.K.M. Martin, D.J. Sorin, B.M. Beckmann, M.R. Marty, M. Xu, A.R. Alameldeen, et al. (2005). "Multifacet's general execution-driven multiprocessor simulator (GEMS) toolset." *ACM SIGARCH Computer Architecture News*, 33(4). DOI: 10.1145/1105734.1105747

B. Prince (1999). *High Performance Memories*. John Wiley & Sons.

R. Sites (1996). "It's the memory, stupid!" *Microprocessor Report*, 10(10).

J. Smith and R. Nair (2005). *Virtual Machines: Versatile Platforms for Systems and Processes*. Morgan Kaufmann Publishers Inc. San Francisco CA, USA.

S. Srinivasan (2007). *Prefetching vs. the Memory System: Optimizations for Multicore Server Platforms*. Ph.D. Thesis, University of Maryland, Dept. of Electrical & Computer Engineering.

M. Talluri, S. Kong, M.D. Hill, and D.A. Patterson (1992). "Tradeoffs in supporting two page sizes." In *Proc. 19th Annual International Symposium on Computer Architecture (ISCA'92)*, pp. 415–424. DOI: 10.1145/139669.140406

M. Talluri and M.D. Hill (1994). "Surpassing the TLB performance of superpages with less operating system support." In *Proc. Sixth Int'l Conf. on Architectural Support for Programming Languages and Operating Systems (ASPLOS'94)*, pp. 171–182. DOI: 10.1145/195473.195531

G. Taylor, P. Davies, and M. Farmwald (1990). "The TLB slice – A low-cost high-speed address translation mechanism." In *Proc. 17th Annual International Symposium on Computer Architecture (ISCA'90)*. DOI: 10.1109/ISCA.1990.134546

N. Tuaycharoen (2006). *Disk Design-Space Exploration in Terms of System-Level Performance, Power, and Energy Consumption*. Ph.D. Thesis, University of Maryland, Dept. of Electrical & Computer Engineering.

D. Wang, B. Ganesh, N. Tuaycharoen, K. Baynes, A. Jaleel, and B. Jacob (2005). "DRAMsim: A memory-system simulator." *ACM SIGARCH Computer Architecture News*, 33(4), 100–107. DOI: 10.1145/1105734.1105748

W.A. Wulf and S.A. McKee (1995). "Hitting the memory wall: Implications of the obvious." *ACM Computer Architecture News*, 23(1), 20–24. DOI: 10.1145/216585.216588

L. Zhao, R. Iyer, J. Moses, R. Illikkal, S. Makineni, and D. Newell (2007). "Exploring large-scale CMP architectures using Manysim." *IEEE Micro*, 27(4), 21–33. DOI: 10.1109/MM.2007.66

Biography

BRUCE JACOB

Bruce Jacob is an Associate Professor and Director of Computer Engineering in the Dept. of Electrical and Computer Engineering at the University of Maryland, College Park. He received his Ars Baccalaureate, cum laude, in Mathematics from Harvard University in 1988, and his M.S. and Ph.D. in Computer Science and Engineering from the University of Michigan in 1995 and 1997, respectively. In addition to his academic credentials, he has extensive experience in industry: he designed real-time embedded applications and real-time embedded telecommunications architectures for two successful Boston-area startup companies: Boston Technology (now part of Comverse Technology) and Priority Call Management (now part of uReach Technologies). At Priority Call Management he was employee number 2, the system architect, and the chief engineer; he built the first working prototype of the company's product, and he built and installed the first actual product as well.

In recognition of Prof. Jacob's research program, he has been honored several times as a University of Maryland "Rainmaker." In memory-systems research, Jacob's cache and memory-management designs demonstrated the viability of software-managed caches in general-purpose systems (he coined the now-common term "software-managed cache" in a 1998 ASPLOS paper). His work in advanced DRAM architectures at Maryland is the first comparative evaluation of today's memory technologies, and he received the CAREER Award from the National Science Foundation for his early work in that area. Honors for his teaching include the departmental George Corcoran Award, the University of Maryland Award for Teaching Excellence, and his 2006 induction as a Clark School of Engineering Keystone Professor. He has published over 50 papers on a wide range of topics, including computer architecture and memory systems, low-power embedded systems, electromagnetic interference and circuit integrity, distributed computing, astrophysics, and algorithmic composition. His recently published book on computer memory systems (Jacob, Ng, and Wang: *Memory Systems—Cache, DRAM, Disk*, Morgan Kaufmann Publishers, Fall 2007) is large enough to choke a small elephant.

Printed in the United States
by Baker & Taylor Publisher Services